Climate Change and Agriculture

NEW HORIZONS IN ENVIRONMENTAL ECONOMICS

Series Editors: Wallace E. Oates, *Professor of Economics, University of Maryland, College Park and University Fellow, Resources for the Future, USA and* Henk Folmer, *Professor of Research Methodology, Groningen University and Professor of General Economics, Wageningen University, The Netherlands*

This important series is designed to make a significant contribution to the development of the principles and practices of environmental economics. It includes both theoretical and empirical work. International in scope, it addresses issues of current and future concern in both East and West and in developed and developing countries.

The main purpose of the series is to create a forum for the publication of high quality work and to show how economic analysis can make a contribution to understanding and resolving the environmental problems confronting the world in the twenty-first century.

Recent titles in the series include:

Climate Change and Agriculture

An Economic Analysis of Global Impacts, Adaptation and Distributional Effects

Robert Mendelsohn

Yale University, USA

and

Ariel Dinar

University of California, Riverside and (during the preparation of this book) World Bank, USA

NEW HORIZONS IN ENVIRONMENTAL ECONOMICS

Edward Elgar
Cheltenham, UK • Northampton, MA, USA

Published by
Edward Elgar Publishing Limited
The Lypiatts
15 Lansdown Road
Cheltenham
Glos GL50 2JA
UK

Edward Elgar Publishing, Inc.
William Pratt House
9 Dewey Court
Northampton
Massachusetts 01060
USA

A catalogue record for this book
is available from the British Library

Library of Congress Control Number: 2009930434

Mixed Sources
Product group from well-managed
forests and other controlled sources
www.fsc.org Cert no. SA-COC-1565
© 1996 Forest Stewardship Council

ISBN 978 1 84720 670 1

Printed and bound by MPG Books Group, UK

Contents

About the authors

Robert Mendelsohn is the Edwin Weyerhaeuser Davis Professor in the School of Forestry and Environmental Studies at Yale University, USA. His research focuses on valuing the environment including air pollution emissions, hazardous waste pollution, wildlife populations, recreation, oil spills, timber, and non-timber forest products. For the last 15 years, Dr Mendelsohn has been measuring the impacts of climate change around the world. He has been involved in studies on agriculture, water, energy, sea level rise and forestry. Dr Mendelsohn has also been especially interested in how society will adapt to future changes in climate.

Ariel Dinar is a Professor of Environmental Economics and Policy, and Director of the Water Science and Policy Center, University of California, Riverside, USA. This book was prepared while he was Lead Economist of the Development Research Group at the World Bank, USA. His research focuses on international water and cooperation, approaches to stable water allocation agreements, water and climate change, climate change and agriculture, economics of water quantity/quality, and economic aspects of policy interventions and institutional reforms.

Acknowledgements

This book compiles work that has been conducted in research funded by the World Bank around the world in the past 15 years. We would like to thank the World Bank Research Committee, The Knowledge for Change Program (KCP), and the Global Environmental Facility (GEF) for funding the various research projects that led to this book. We bow in recognition of the enormous efforts made in data collection and the dedication of the country teams in Africa and Latin America which allowed us to undertake the analysis that forms the basis of this book. We would like to thank The Centre for Environmental Economics and Policy in Africa (CEEPA) and PROCISUR for their coordination of the country work in Africa and South America. This book could not have been written without the hard work of Apurva Sanghi, Pradeep Kurukulasuriya and Niggol Seo, all former Ph.D. Students at Yale who found climate change and agriculture to be the focus not only of their Ph.D. thesis, but also of their future careers. Special appreciation is due to Polly Means and Emanuele Massetti for their work on producing the maps that feature in the book. And finally, we benefited from very useful comments from Richard Adams and Lewis Ziska.

1. Introduction

This book examines the impact of climate change on agriculture and what farmers do to adapt to climate. Agriculture and grazing currently occupy 40 percent of the earth's land surface (Easterling and Aggarwal et al., 2007). Overall, agriculture is responsible for 6 percent of the world's GDP. In many developing countries, agriculture is a primary sector of the economy and is the primary source of livelihood for about 70 percent of the rural population (Easterling and Aggarwal et al., 2007). Climate changes are expected to affect farmers throughout the world. This book examines the magnitude of the impacts to farmers, where these impacts will occur, and how farmers can adapt.

Although climate change is expected to have many impacts on various sectors, few sectors are as important as agriculture. If future climate scenarios lead to a widespread reduction in food supply, there could be massive problems with hunger and starvation (Rosenzweig and Parry, 1994; Reilly, 1996). Global analyses of the total impacts of rising greenhouse gases have consistently raised concerns about agricultural impacts (Cline, 1992; Pearce et al., 1996; Reilly et al., 1996; Gitay et al., 2001; Easterling and Aggarwal et al., 2007). Virtually all developed countries are concerned about whether climate change will damage their agricultural sectors. However, several authors are concerned that agricultural losses will be especially harmful to developing countries (Pearce et al., 1996; Rosenzweig and Parry, 1994; Mendelsohn and Williams, 2004; Cline, 2007).

CLIMATE CHANGE

In order to understand climate impacts, it is first necessary to discuss climate change itself. It is now widely agreed that changes in land use and especially burning fossil fuels have already caused and will continue to cause substantial releases of greenhouse gases into the atmosphere (IPCC, 2007). These emissions stay in the atmosphere for long periods of time so that the concentrations of greenhouse gases have been rising steadily (IPCC, 2007). As greenhouse gas concentrations rise, they are expected to trap heat in the lower atmosphere. This excess heat is expected to warm

Table 1.1   Projected global average surface warming at the end of the 21st century

Scenario[a]	Temperature change[b] (change in °C in 2090–2099 compared to 1980–1999)	
	Best estimate	Likely range
A1T (600 ppm)	+2.4	1.4–3.8
A1B (800 ppm)	+2.8	1.7–4.4
A2 (1250 ppm)	+3.4	2.0–5.4
A1F1 (1550 ppm)	+4.0	2.4–6.4

Notes:
a.   The scenarios listed above reflect likely outcomes in the absence of mitigation over the next century.
b.   To express the temperature changes relative to pre-industrial times add 0.5°C.

Source: Adapted from IPCC (2007, Table SPM3).

the oceans over several decades. The warmer oceans in turn lead to a long-term change in climate. Because it takes oceans so long to warm, there is a lag between emissions and temperature changes.

There is evidence that temperatures have warmed about 0.5°C over the past century (IPCC, 2007). The climate change of concern, however, is not past warming but rather warming in the future. If there is no mitigation of greenhouse gas emissions over the next century, global temperatures are expected to rise between 2°C and 4°C depending on the emissions scenario (IPCC, 2007). However, even these estimates are uncertain, so the range of actual warming by 2100 may be even broader. One reason why the range is so wide is that it is not clear how much greenhouse gas the future economy will emit. A second reason is that it is not clear how much CO_2 will be absorbed by the biosphere and the ocean. A third reason is that it is not clear whether other forces such as sea ice and clouds in the earth-climate system will dampen or enhance the greenhouse effect.

Table 1.1 examines a number of likely scenarios for the next century if there is no mitigation. It is quite clear that the higher the concentrations, the greater the warming. However, even with a known level of green-house gas concentrations, there is still uncertainty about the magnitude of warming. The direct effect of man-made greenhouse gase emissions on solar radiation is reasonably well understood and calibrated. This direct effect is small and leads to the lower estimates in the range. In addition, climate scientists expect that there are positive feedbacks. Once climate changes, these positive feedbacks can amplify the change in climate from man-made emissions (IPCC, 2007). For example, warming will increase the speed of the hydrological cycle and there is likely to be increased cloud. These clouds could themselves act as greenhouse gases. However, depending on the height at which they form, the clouds may act as a cooling force. Sea ice will melt as temperatures rise and this will change the albedo of the earth's surface so that it absorbs more heat. However, it is not clear how powerful this effect will be at the poles where such an effect will occur. The albedo of other parts of the earth's surface may also change as ecosystems shift. There may also be releases of carbon from the biosphere as it warms. All these effects have the potential to amplify man-made emissions into much larger changes in temperature, but they are uncertain.

It is also important to note that the anticipated temperature increases are likely to be greater near the poles and lower near the equator. The temperature changes will not be uniform across the planet. Further, the changes in temperature are likely to increase the energy in the hydrological cycle, leading to more clouds, more rain and more evaporation. These changes can be as important as the change in temperature. How precipitation changes will be distributed across the planet is not clear as meteorological processes are likely to change, leading to weather patterns shifting from place to place. Some areas may get a lot more rain and others a lot less. These changes are poorly understood at the moment and there is little agreement about the distribution of precipitation changes across the planet across climate models.

In conclusion, there is a great deal of uncertainty about future climate change scenarios. The question is not whether greenhouse gases might warm the planet; the answer to this question is clearly that they will. The issue is how much will the planet warm, how much will precipitation change, and how will these changes be distributed across the planet? This book does not answer these questions. These answers lie in climate science.

The focus of this book is upon measuring the consequences of climate changes for agriculture. In order to capture the uncertainty in climate outcomes, the book examines the range of climate changes currently considered plausible. In particular, the book contrasts a relatively mild warming scenario with a relatively harsh scenario. The mild scenario predicts a small increase in temperature (of about 2.5°C) and a small increase in precipitation (of about 7 percent). The harsh scenario predicts a large increase in temperature of about 5°C and no increase in precipitation. There is a myriad of possible climate scenarios but examining these two provides a reader with a sense of the consequences of the range of possible climate outcomes.

Although climate change is sometimes discussed as though it is a discrete event (before or after), it is actually a continuous process that evolves

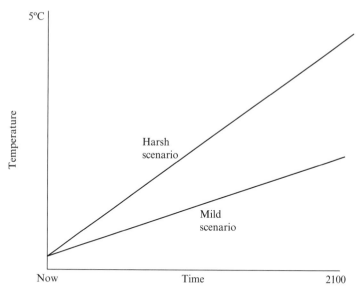

Source: Authors.

Figure 1.1   Future climate scenarios

over time. If emissions are not curbed, temperatures will continue to
increase for centuries. Just looking over the rest of this century indicates
a clear dynamic process. Figure 1.1 illustrates how temperature might
evolve over the rest of this century, depending on whether the outcome
is relatively mild or severe. The differences between the two scenarios are
not that large at first. However, after 2050, the high and low scenarios
become more distinct. By 2100, the two scenarios lead to very different
temperature outcomes.

Climate scientists predict that there may be other changes to climate
in addition to just a gradual increase in annual temperature or precipitation. Warming may change seasonal distributions. Warming may reduce
diurnal range by reducing heat loss at night. Warming may lead to an
increase in interannual variance. Although not every study examined
each of these elements, at least some studies were able to address them all.
The changes above are covered as part of the analysis of climate change
impacts on agriculture in this book.

In addition to the climate changes listed above, scientists also predict
that greenhouse gas emissions will lead to sea level rise, which will
inundate coastal properties. Warming may also increase the severity or
frequency of hurricanes and other storms. These latter two effects are

not covered in this book. That is not to say that these impacts are not important, but that they are quantified as separate studies of impacts. For example, there is a whole collection of literature on sea level rise (Bosello et al., 2007; Yohe et al., 1999; Neumann and Livesay, 2001; Ng and Mendelsohn, 2005; 2006). In addition, there is a small but important body of literature on hurricane and storm damages (West and Dowlatabadi, 1998; Nordhaus, 2006).

ORGANIZATION OF THE BOOK

Chapter 2 establishes the scientific relationship between climate and agricultural production, explaining why plants and animals are sensitive to climate. The chapter explains why agricultural production is affected by climate. However, it also discusses the fact that agricultural performance is also sensitive to soils, technology, institutions and market access.

Chapter 3 reviews the economic literature that has examined climate change impacts on agriculture. A wide range of techniques have been developed including (1) crop modeling; (2) mathematical programming; (3) Ricardian cross-sectional analysis; (4) analysis of panel data; and (5) other econometric approaches. The advantages and disadvantages of each of these approaches are discussed, including how well each approach deals with adaptation. Although there are now many available approaches to studying impacts on agriculture, the bulk of the work has focused on the United States. There has been surprisingly little empirical economic research given the importance of agricultural impacts and especially the importance of agricultural impacts in developing countries.

The remainder of the book reviews a new set of economic studies that measure the impact of climate change and adaptation on agriculture, specifically focusing on developing countries. Chapter 4 develops in detail the impact methodology, the Ricardian approach, used in many of these studies. The Ricardian approach is at the heart of the seven studies funded by the World Bank in four continents, covering 22 countries. The chapter discusses how the Ricardian approach has evolved over time to address different situations and circumstances. The chapter reviews various modifications and adjustments that have been made in data collection and analytical methods.

Chapter 5 examines methodologies for measuring adaptation. Although many authors have been writing about climate adaptation in the abstract and about concrete steps regarding how to adapt to weather, there are very few adaptation studies that show how farmers would adapt to climate change. This chapter discusses new cross-sectional methods to measure how farmers alter their choices in response to climate. The studies examine such choices as whether to grow crops or raise livestock, whether to irrigate, which crops to plant, and which livestock to own. The research quantifies how these decisions are affected by climate.

Chapter 6 explores the merging of adaptation and Ricardian impacts into structural farm models. As with the two earlier approaches discussed in Chapters 4 and 5, this chapter relies on cross-sectional analysis. The structural approach begins with climate, soils and other exogenous variables and ends with predictions of net revenue, as in the Ricardian approach. However, rather than treating adaptation as a black box, the modeling explicitly captures the many choices that farmers make in response to climate, as in the adaptation chapter. The model makes clear what changes farmers would make on a farm that faces a different climate. The models then estimate the conditional income that a farmer would receive, given his/her choices. As a result, the approach can estimate the final expected income of a farmer and can predict how expected income would change as climate changes.

Chapter 7 reviews specific empirical studies that were conducted in the United States, Brazil, India and Sri Lanka, which relied on secondary data in each of those countries. The chapter discusses specific problems that had to be overcome in order to perform the analyses. For example, one important issue was how to measure farm performance. Brazil and the United States have land value data that can measure long-term farm net revenue, but such data does not exist in India. The Indian study consequently had to rely on farm net revenue. Another key question that is discussed is how to measure climate. Weather stations are scarce in rural parts of developing countries and yet there are serious drawbacks to using satellite data. Finally, the chapter also reviews the results from the countries studied. Although there are qualitative results that are consistent across all the countries, the quantitative results are quite different.

Chapter 8 discusses the application of the Ricardian method to Africa, South America and China. The continental scale analyses relied on primary data that was collected specifically for the study. Although the African and South American studies were designed to be as similar as possible, the studies had to make adjustments for each continent. In Africa and China, land value data is not available, so these studies relied on calculated net revenues to measure farm performance. In contrast, the South American data had reliable measures of land values. In Africa, most farmers relied on common lands to graze their livestock so it was not possible to measure the land used by livestock management. In South America, livestock management was largely done on farm and so was captured by farm land value. Satellite data was used to measure temperature in these studies and weather station data was used to measure precipitation. In addition, the African data had measures of water flow through the district. Chapter 8 also reviews the results for each of these studies, which were quite distinct for each region.

Chapter 9 discusses empirical studies of adaptation that were undertaken in Africa and South America. Several decisions by farmers were found to be sensitive to climate: whether or not to plant crops, raise animals or do both depends on climate. Which crops to plant and which livestock to raise are also climate-sensitive decisions. Finally, whether or not to use irrigation was found to depend on both climate and water.

Chapter 10 estimates several structural Ricardian models. The first empirical model begins with the decision to adopt irrigation and then computes the conditional income of dryland and irrigated land. All three equations are climate-sensitive. The second model begins with the decision regarding whether to have crops, livestock or both. For farmers that choose crops, they must then choose whether or not to use irrigation. Finally, for every type of farm, the model estimates conditional income. Both of these farm models predict the impact of climate change. However, in contrast to traditional Ricardian models, these structural Ricardian models also explain how farmers change their behavior.

Chapter 11 synthesizes the results over the 22 countries and four continents. The review examines the overall impacts of climate change on agriculture. The empirical studies consistently reveal that climate is important. In locations that are relatively cool, warming has only a small impact on farm outcomes. But in places that are already hot, warming significantly reduces net revenue and land values. The effect of precipitation changes depends greatly on the level of existing precipitation. In places that are relatively dry, increased precipitation is highly beneficial. However, in places that currently receive relatively heavy amounts of precipitation, increased precipitation is harmful. Rainfed farms are more sensitive to temperature and precipitation changes than irrigated farms. In fact, in several places, warming increases the net revenue of irrigated farms. Provided there is adequate water, irrigation may be an effective adaptation to climate change. Another effective adaptation is shifting from crops to livestock. Shifting crops and shifting livestock species are also very effective climate adaptations. Contrary to expectations, many household farms may readily adapt to climate change because they have good substitutes for what they are doing now. Larger, more commercial farms may be more vulnerable because they have specialized in crops and livestock that are profitable but more heat-sensitive.

Chapter 12 draws overall policy implications at various levels, including local, country, regional and global levels. Because impacts vary significantly across space, the actual impact on each farm is a local phenomenon. Consequently, adaptation must also be local. Some barriers to adaptation, however, are at the national level. For example, moving resources from public and common property ownership to private property ownership increases efficient adaptation. Finally, some adaptations may be regional or global. Encouraging free trade of farm products across climate zones increases the overall resilience of agriculture so that people become less dependent on local productivity. Developing more suitable crop varieties and animal breeds for a warmer world is more effective if there is a global market for the products rather than a national market. Other issues such as improved public management and income transfers are also addressed.

Chapter 12 also addresses future research needs. The chapter discusses the need to improve data collection, extend the analyses to all regions, and improve the analytical methods. One especially critical area that needs to be developed is closer interaction between agriculture and other sectors (especially water and forests).

REFERENCES

Bosello, F., R. Roson and R.S.J. Tol (2007), 'Economy-wide estimates of the implications of climate change: sea level rise', *Environmental and Resource Economics*, **37**, 549–71.

Cline, W.R. (1992), *The Economics of Global Warming*, Washington DC: Institute for International Economics.

Cline, W.R. (2007), *Global Warming and Agriculture: Impact Estimates by Country*, Washington, DC: Center for Global Development and Peterson Institute for International Economics.

Easterling, W. and P. Aggarwal et al. (2007), 'Food, fibre, and forest products', in IPCC, *Climate Change 2007: Impacts, Adaptation and Vulnerability*, Fourth Assessment Report, Cambridge, UK: Cambridge University Press, pp. 275–313.

Gitay, H., S. Brown, W. Easterling and B. Jallow (2001), 'Ecosystems and their goods and services', in IPCC, *Climate Change 2001: Impacts, Adaptation, and Vulnerability*, Cambridge, UK: Cambridge University Press, pp. 235–342.

IPCC (2007), *Climate Change 2007: The Physical Science Basis*, contribution of Working Group I to the Fourth Assessment Report of the Intergovernmental Panel on Climate Change, Cambridge, UK: Cambridge University Press.

Mendelsohn, R. and L. Williams (2004), 'Comparing forecasts of the global impacts of climate change', *Mitigation and Adaptation Strategies for Global Change*, **9**, 315–33.

Neumann, J. and N. Livesay (2001), 'Coastal structures: dynamic economic modeling', in R. Mendelsohn (ed.), *Global Warming and the American Economy: A Regional Analysis*, Cheltenham, UK and Northampton, MA, USA: Edward Elgar, pp. 132–48.

Ng, W. and R. Mendelsohn (2005), 'The impact of sea level rise on Singapore', *Environment and Development Economics*, **10**, 201–15.

Ng, W. and R. Mendelsohn (2006), 'The impact of sea-level rise on non-market lands in Singapore', *Ambio*, **35**, 289–96.

Nordhaus, W. (2006), 'The economics of hurricanes in the United States', http://nordhaus.econ.yale.edu/cv_current.htm.

Pearce, D., W. Cline, A. Achanta, S. Fankhauser, R. Pachauri, R. Tol and P. Vellinga (1996), 'The social cost of climate change: Greenhouse damage and the benefits of control', in IPCC, *Climate Change 1995: Economic and Social Dimensions of Climate Change*, Cambridge, UK: Cambridge University Press, pp.179–224.

Reilly, J. et al. (1996), 'Agriculture in a changing climate: impacts and adaptations', in IPCC *Climate Change 1995: Impacts, Adaptations and Mitigation of Climate Change*, Cambridge, UK: Cambridge University Press, pp. 427–68.

Rosenzweig, C. and M. Parry (1994), 'Potential impact of climate change on world food supply', *Nature*, **367**: 133–8.

West, J.J. and H. Dowlatabadi (1998), 'On assessing the economic impacts of sea level rise on developed coasts', in T.E. Downing, A.A. Olsthoorn and R.S.J. Tol (eds), *Climate, Change and Risk*, London: Routledge, pp. 205–20.

Yohe, G., J. Neumann and P. Marshall (1999), 'The economic damage induced by sea level rise in the United States', in R. Mendelsohn and J. Neumann (eds), *The Impact of Climate Change on the United States Economy*, Cambridge, UK: Cambridge University Press, pp. 178–208.

2. The role of climate in agricultural production[1]

In this chapter we discuss the natural scientific mechanisms through which climatic variables affect crops and livestock. Crops and livestock are affected by climate via multiple mechanisms, directly and indirectly. The natural science literature addressing the physiology of crops and animals is quite complicated and yet still incomplete (Wolfe and Erickson, 1994, p. 154; NRC, 1981, p. 1). However, there is sufficient evidence to expect that climate change will affect agricultural productivity.

The chapter discusses how each climatic effect interacts with crop production and animal growth and production. First, both controlled laboratory experiments and field experience reveal that each crop or animal has an optimal range of climate in which that crop or animal yields the highest growth and production. Sub-optimal climate conditions lead to lower growth rates and production levels. Second, various factors directly and indirectly affect the ability of crops and livestock to grow and produce in a given climate. Background factors such as soils and water may interact with climate and change the relationship of climate and production. For example, a poor soil may not only reduce overall production, but it may make a crop more sensitive to temperature or precipitation as well. A stylized depiction is shown in Figure 2.1. With good soil and no other limitation, growth and production are maximized in the optimal climate range. With poor soil, growth and production are lower in the optimal climate. However, the interaction between the poor soil and higher temperature is even more harmful. The crop does even less well if faced with both a temperature that is too high and a soil that is poor. A similar interaction holds between a harsh climate and poor grazing conditions for livestock.

CROP GROWTH AND PRODUCTION WITH CLIMATE CHANGE

In order to understand the mechanisms affecting crops' responses to climatic factors, we start by describing plant physiology. For crops to grow

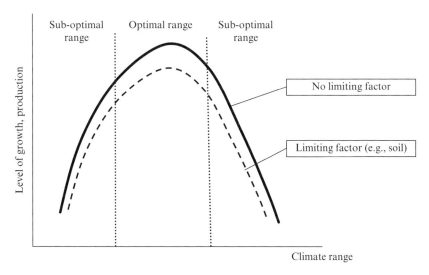

Source: Authors.

Figure 2.1 *Impact of climate and soil interaction on growth and production of crops and livestock*

and be economically productive they have to have optimal amounts of various inputs such as sunlight, water, carbon dioxide (CO_2), nutrients, and limited weeds, disease and insects (Wolfe and Erickson, 1994). For growth and production, photosynthesis converts the energy of sunlight into chemical potential energy stored in the organic structures of plants. The process is very complex but, put simply, the plant uses sunlight to convert CO_2, water, nitrogen (N), and other resources to oxygen (O_2), water and carbohydrates. The carbohydrates are then used for growth and energy (Ke, 2001). The relative distribution of dry matter among the plant's organs (for example, roots, stems, leaves, flowers[2]), determines the efficiency of the growth process[3] relative to economic yield (Morison, 1996). The level of CO_2 in the atmosphere per se is a factor affecting plant growth and yield directly and indirectly. The direct effect of CO_2 rise is to increase photosynthesis and growth. The indirect effect of rising CO_2 is that it reduces plant water loss. Temperature and humidity also play important roles in this process. Plants have become specialized to optimize yields at particular temperature and humidity settings (Morison, 1996). Combinations of temperature and CO_2 levels affect the rate of development of plants (Figure 2.2). Under conditions of climate change, elevated CO_2 and temperature levels increase the rate of plant development, thus

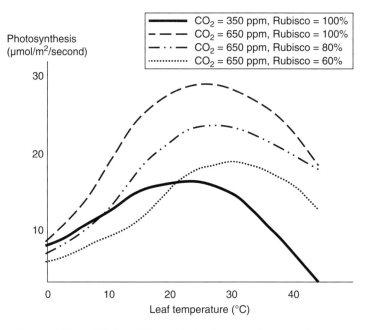

Note: Grace and Zhang (2006, p. 196) provide simulation results that are in agreement with the trends in this figure.

Source: Based on Wolfe and Erickson (1994, p. 163, Fig. 8-2).

Figure 2.2   *Photosynthesis per unit of leaf area as affected by CO$_2$ concentration and activity level of the key photosynthetic enzyme, Rubisco*

shortening the growing period. But, this may result in less potential growth and thus reduce the overall yield production (Morison, 1996).

As can be seen from Figure 2.2, photosynthesis–temperature relations are hill-shaped. The control is the process under CO$_2$ concentration of 350 ppm and full activity of the photosynthesis Rubisco enzyme. The CO$_2$ benefits are much less apparent in a temperature range below 20°C for the two levels of reduced activity of the photosynthesis Rubisco enzyme.[4] At higher temperatures, the photosynthesis process is more effective for higher concentrations of CO$_2$ than in the control case for all level of Rubisco's activity (Wolfe and Erickson, 1994). For reasons that are not clear, only the FACE (Free Air Carbon Dioxide Enrichment) experiments lead to the results reported in Figure 2.2. There is very little evidence for CO$_2$–temperature interactions under field conditions.

Increased temperature implies increased demand for water, but elevated levels of CO_2 concentration lead to reduced stomata openness and thus to reduced water requirements by plants. This leads to more water efficient yield production. The nutrient–CO_2 interactions are less clear (Morison, 1996).

There are numerous studies that have shown the effects of projected and current increases in CO_2 on various aspects of crop growth and yield (see summary in USDA and USCCSP, 2008). Overall, these can be divided into two categories. Those crops with the C4 photosynthetic pathway (mostly tropical grasses) are already optimized at the current CO_2 level. They will have only a small response to higher CO_2. Crops with the C3 photosynthetic pathway (approximately 95 percent of all crop species) will respond the most to rising CO_2 levels.

However, to fully exploit any potential gain from rising CO_2 levels, it is necessary to determine how other environmental variables influence any positive response. Among such variables, temperature is projected to rise with levels of atmospheric CO_2. Potentially, at the biochemical level, there could be a beneficial interaction of CO_2 enrichment and temperature on dry matter production in C3 crops due to the enzymatic characteristics of the Rubisco enzyme. For example, increasing temperature and carbon dioxide could be beneficial for production of lettuce or spinach, because yields for these plants are defined as total above ground dry matter. In contrast, there do not appear to be any synergistic effects between CO_2 and temperature in reproductive (i.e., floral) yields (USDA and USCCSP, 2008), and, in some cases, a negative effect is observed between the relative enhancement of yield with increasing CO_2 and rising temperature. Such effects have been observed for rice (Kim et al., 1996; Matsui et al., 1997), sorghum (Prasad et al., 2006) and dry bean (Prasad et al., 2002).

Soil Moisture and Nutrients

Two of the most important edaphic factors related to crop yields are soil moisture and nutrient availability. For rice and wheat, there is now sufficient data to indicate that any stimulation of crop yield by increasing CO_2 will be dependent on N availability (for example, see Kim et al., 2003 for rice; Wolf, 1996 for wheat) whereas stimulation of yield by CO_2 in soybean appears independent of supplemental N (Cure et al., 1988). However, a number of questions remain unaddressed.

In contrast to nutrients, there is a surfeit of data indicating that under water limiting conditions, the indirect effect of CO_2 on stomatal aperture (and potential reductions in transpirational water use) may enhance the relative effect of elevated CO_2 on crops (see Polley, 2002 for a review). Irrigation can allow some crops to grow in places where they would not

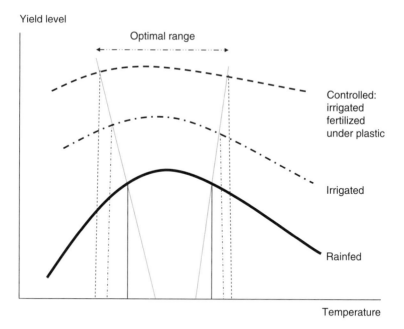

Source: Authors.

Figure 2.3   Schematic presentation of temperature–yield relationship with various enabling technologies

naturally prosper. Specifically, irrigation has been used to grow crops in places that are too dry to support plant growth. However, irrigation also reduces plant sensitivity to temperature. Figure 2.3 describes the relationship between yields and temperature for rainfed crops in the lower curve. If the farmer shifts to irrigation, not only do yields increase overall, but the sensitivity to temperature flattens. If the farmer then installs covers, the sensitivity to temperature declines even further. By investing in these adaptation strategies, the farmer can cope with sub-optimal climate conditions. Of course, these investments are costly and so the farmers must always be aware of the net gains. One important aspect of water availability is its temporal variability over the growing period. As argued by Pereira et al. (2006), plant productivity is higher if the water supply is continuous and regular as opposed to spiky and irregular, depending on the growth stage.

Recent work based on experiments with FACE suggests lower CO_2 fertilization effects than past estimates, which relied on closed chambers at small scales to simulate open-air field conditions. With CO_2 concentrations of 550–575 ppm, the FACE experiments show that the 'yield increase is 11% for C3 crops and 7% for all five major food crops' (Long et al., 2005). This is one-third to one-quarter of the direct effect of CO_2 modeled in the recent assessment for Europe and the USA (Darwin and Kennedy, 2000). In a more recent study, Long et al. (2006) find a 550 ppm carbon dioxide concentration leads to a 13 percent increase for wheat in contrast to the 31 percent increase found in enclosure studies, a 14 percent increase instead of a 32 percent increase for soybeans, and no increase instead of an 18 percent increase for C4 crops (Long et al. 2006). However, these FACE experimental results remain controversial.[5]

In response to the FACE criticism, another study extends the analysis of plant response to rising CO_2 levels from the cell and leaf levels to the whole plant response (Ziska and Bunce, 2006). CO_2 may shift biomass allocation in the individual plant, either increasing or decreasing productivity. This can significantly alter the results of CO_2 impacts calculated from controlled laboratory experiments under optimal growing conditions.

The scientific experiments available in the literature are limited in their capacity to incorporate interactions with various factors. The CO_2–temperature impact on plant photosynthesis and on yield needs to be evaluated further with respect to the availability of other limiting production factors such as water, nutrients, soil quality and pollution. Crop models may or may not take into account these many interactions depending upon their assumptions. In the next section we address the principles of interaction between several essential biotic factors of production such as diseases, pests and weeds.

Pathogens and Diseases

CO_2 and climate change may have not only a direct but also an indirect effect on crop plant yields via disruption of pest dynamics, specifically changes in pathogens, insects and weeds. The relationship between climate variation, particularly temperature and water availability, and the incidence and severity of plant pathogens and disease has long been recognized (Colhoun, 1973). Mild winters and warmer weather have been associated with increased outbreaks of powdery mildew, leaf spot disease, leaf rust, and rizomania disease (see Patterson et al., 1999). Presumably in part, these effects of warmer winters are caused by an increase in the amount of inoculum present in the spring. Warm, humid conditions may result in earlier and stronger incidence of late potato blight (*Phytophthora infestans*), a devastating disease of significant historical importance (Parry et al., 1990). Consistently warmer temperatures would also be likely to shift the range of diseases into cooler regions (Treharne, 1989).

Increased precipitation is likely to increase the spread of diseases since rain and splash water both spread spores (Royle et al., 1986) and wet plant surfaces are necessary for spore germination and infection to proceed. Conversely, increasing aridity could lessen disease problems, although some diseases such as powdery mildew are promoted by hot, dry daytime conditions if nighttime temperatures result in dew formation (Ziska and Runion, 2007). While extreme climatic conditions (for example drought, flood) will undoubtedly impact micro-organisms directly, their effects on the stress to host plants is also of concern. Stressed plants, such as those suffering from drought, are often more susceptible to pathogen and abiotic attack. Such stresses are often cited as primary contributors to outbreaks of disease such as diebacks and declines (Manion and LaChance, 1992).

Overall, the projected changes in temperature, water and extreme weather events are likely to influence the range of attacking pathogens and lead to changes in outbreaks of new and existing plant diseases. But, as yet, there is not enough information to predict what will happen.

Weeds

As with crops, weeds are also likely to be directly stimulated by rising CO_2 levels. Although a number of weeds do have the C4 photosynthetic pathway and would not be expected to respond strongly to rising CO_2 levels, crop–weed associations are complex, and for many crops the 'worst' weed is simply a wild relative that has adapted to the cultural and environmental habitat of its domesticated cousin. Field studies that have examined the impact of weeds on crop yield with the same photosynthetic pathway have found that weeds are universally favored by increasing CO_2 (Ziska and Bunce, 2006).

Pests

Precipitation extremes such as droughts or floods are associated with changes in insect herbivory and these changes will have significant impacts on agricultural ecosystems (Fuhrer, 2003). As with temperature, projected changes in extreme precipitation events are likely to shift the occurrence and frequency of insect outbreaks. Overall, there is initial evidence suggesting a strong change in altitudinal and longitudinal migrations for a number of insect pests, and a greater rate of feeding with rising CO_2 and climate change. Although high temperature stress could increase crop vulnerability to insects directly, temperature is also widely recognized as the principal abiotic factor controlling insect growth and development.

Although weed impacts are likely to be exacerbated with CO_2 and climate, it is also clear that weed management, particularly chemical weed management, is also likely to be impacted. There are an increasing number of studies (for example, Ziska et al., 1999; Ziska and Teasdale, 2000) that demonstrate a decline in pesticide efficacy with rising CO_2.

RESPONSE OF SEVERAL CROPS TO CLIMATE CHANGE

Important crop plants have been studied across various climates for many years. We provide results of a couple of these studies for cotton (Figure 2.4) (Reddy et al., 2000) and rice (Figure 2.5) (Hartwell, 2008). More information on maize and sorghum can be found in Young and Long (2000); on root and tuber crops in Miglietta et al. (2000); on tree crops in Janssens et al. (2000); and on productive grasslands (Nosberger et al., 2000).

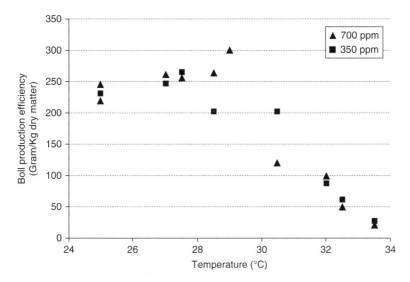

Source: Adapted from Reddy et al. (2000, p. 180, Fig. 81).

Figure 2.4   Cotton boll production efficiency by temperature and CO_2 levels

Wheat

Wheat is an important staple food and has been studied thoroughly. However, experts are still uncertain about the mechanism by which photosynthesis occurs in wheat (Lawlor and Mitchell, 2000, p. 63). The impact of temperature on wheat growth and production in the presence of CO_2 is sigmoidal rather than linear, with a growth range starting at 1–5°C, and that rises slowly and reaches its maximum at 30°C, after which it rapidly slows down (Lawlor and Mitchell, 2000, p. 65). Wheat production is very sensitive to the impact of heat and other factors during the reproductive periods.

Most wheat production takes place in dry regions. However, water is a crucial factor in wheat production. The interaction between elevated CO_2 levels and water (water-stress, drought) suggests that water use by wheat may decrease slightly under wet conditions, but may increase slightly under dry conditions; and that the stimulation of biomass and grain yields tend to be greater under drought than wet conditions (Lawlor and Mitchell, 2000, p. 69).

Rice

Rice contributes to a major share of the dietary requirement of one-third of the world's population. Studies predict that climate change may have a substantial positive or negative impact on rice production, depending on the region (Horie et al., 2000). Rice growth is very sensitive to temperature. Different ranges of temperature during various growth stages (mainly flowering) reduce yield and may end up in different vegetative vs. reproductive outcomes (Horie et al., 2000, p. 84).

Rice displays higher crop water-use efficiency (WUE) under elevated levels of CO_2. This is mainly because of increased biomass production and partly due to reduced transpiration (Horie et al., 2000). However, warming beyond the 24–26°C range reduces WUE sharply.

Several different studies suggest that doubling CO_2 increases rice biomass production under field conditions by about 24 percent when there is sufficient soil N. With appropriate fertilization, biomass production and CO_2 enhancement may reach an additional 30 percent (Horie et al., 2000, p. 100). The results in Figure 2.5 are based on a controlled experiment (Hartwell, 2008) exploring the interaction between temperature and CO_2 impact on the yield of rice with no other factor limitation. They suggest similar trends to those in Horie et al. (2000). Figure 2.5 also supports the hill-shaped relationship between yield and temperature.

ADAPTATIONS RECOMMENDED FOR CROPS

Given the documented and projected changes in CO_2 and climate on crop yields and variability, there is an impetus for adapting crops in order to maintain food supply and food security. The concept of adaptation to

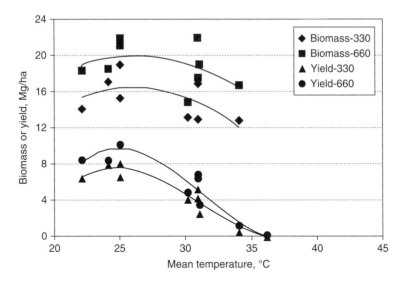

Figure 2.5   *Response of Indica rice (variety IR30) to doubling in CO_2 for a range of likely air temperatures*

climate change covers several aspects, including management changes by the farmers and changes to crop genotype to allow better performance of the crops under changing climate conditions. Crop scientists have recommended the following adaptations.

Sowing Days

The interaction between elevated levels of CO_2 and increased temperature, with adequate water and soils, may lead to shortening of the growing season (Wahaj et al., 2007). One possible adaptation option is to adjust the sowing dates. There is a significant variation in climate (temperature) in different sites in eastern India during the growing season of rice (Krishnan et al., 2007). The study found that the effect of changing sowing date on the yield of rice was significant in several locations. These insights can be used to predict the effect of changing sowing dates with different future climate scenarios. For example, changes to rice yields (variety IR36) under three Global Circulation Models (GCMs) (GFDL, GISS, UKMO) are presented in Table 2.1 for five sowing dates in Jorhat, India.

Changes to sowing dates may necessitate other management adjustments as well (Baker and Allen, 1993 and Baker et al., 1992a, b).

Table 2.1   Change in rice yield (%) across future climate scenarios under various sowing dates

GCM	Sowing date				
	1 June	15 June	1 July	15 July	1 August
GFDL	−9.4	13.5	27.1	17.2	−6.8
GISS	−8.6	12.3	24.3	16.8	−7.5
UKMO	−17.5	7.6	13.4	5.4	−16.4

Source: Jorhat, India. Adapted from Krishnan et al. (2007).

Breeding Strategies and Biotechnology

Some of the difficulties imposed by climate change on crop growth and production could be addressed by proper breeding to allow production in climates they have not been exposed to. Hall and Ziska (2000) and Newton and Edwards (2007) suggest that exploitation of intraspecific variability among different plant lines could enhance productivity affected by rising CO_2 levels. Suggested breeding strategies also include: selecting for higher efficiency in nutrient absorption rate; improved water stress and salinity tolerance (tomato); ability to set seeds and fruits under hot conditions (tomato, Pima cotton); and better crop resistance to pests. This last objective is the most difficult breeding option.

Biotechnology may address at least several of the major stresses crop plants face from direct and indirect climate change impacts (Cheikh et al., 2000). For example, salinity and drought tolerance have been genetically engineered in tobacco, maize, barley and rice. Tolerance to low temperatures has been engineered in tobacco, expanding its range. Crops that have developed drought or salinity tolerance also have tolerance to sub-freezing temperatures (e.g., potato). Less effort was put into developing genetic modifications to allow heat tolerance in crops, mainly because heat stress was associated with drought, which has been widely addressed (Cheikh et al., 2000).

ANIMAL GROWTH AND PRODUCTION WITH CLIMATE CHANGE

Livestock are used for producing certain food products and services for human consumption. Livestock's metabolism is designed to convert feed energy both for maintaining body heat and for production of meat, milk, eggs, wool and other products (NRC, 1981). In many developed countries,

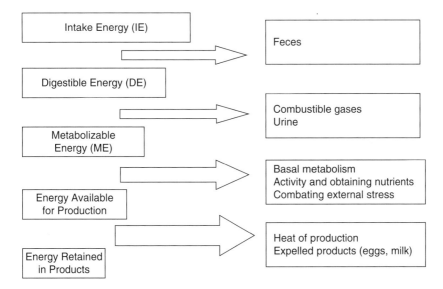

Source: Based on NRC (1981, Fig. 3).

Figure 2.6 The process of feed energy conversion in animals

livestock husbandry takes place in optimal controlled settings. Animals live in shelters, are provided with water on demand, and are given feed for nutrients. However, in many developing countries, animals live outside and depend on natural conditions for both water and food. Animals in developing countries are therefore much more sensitive to environmental factors such as climate. Climate can affect the animals directly, for example in droughts or cold freezes, or indirectly by affecting local plant productivity and ecosystems and therefore nutrient inputs. Environmental conditions can also affect pests, leading to disease vectors. As with plants, different species of animals are adapted to live in different ranges of climate conditions. Management practices such as providing supplemental water or feed can address sub-optimal environmental conditions, especially in the short run.

Animal production and growth are sometimes criticized because they require more energy per calorie of product than plants. The process that converts feed intake to energy and products under environmental stress is presented in Figure 2.6.

Figure 2.6 shows that there are many steps from initial feed energy to final products. In the process of growing animals, some primary energy is lost. However, from the perspective of human consumption, the result is

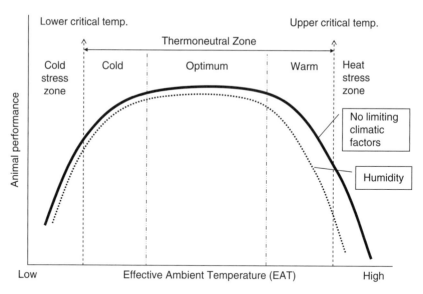

Source: Based partly on NRC (1981, p. 9).

Figure 2.7   Relationship between animal productivity and temperatures

a wide variety of highly valued products that cannot be obtained directly from plants.

The primary determinant of the thermal environment for farm livestock is air temperature. The Effective Ambient Temperature (EAT) is a measure of the thermal environment (NRC, 1981). Each species has an optimal EAT range. However, livestock can compensate for sub-optimal EAT levels by changing food intake, metabolism and heat dissipation at least to a limited extent (NRC, 1981). Radiation, humidity, air movement and precipitation determine how EAT affects animal performance.

In a similar way, as in crops, we can draw a relationship between climatic conditions, measured in terms of EAT, and animal health and performance (Figure 2.7). We can verify the above theoretical relationship experimentally. For example, there is a relationship between milk production and heat stress (Finocchiaro et al., 2005). The Temperature–Humidity Index (THI) was used as an indicator for the degree of stress on animals caused by temperature and humidity. The THI was calculated by combining maximum temperature, T, (in °C) and average relative humidity, RH, (%) using the following expression (temperature varied between 24 and 33°C and relative humidity varied between 50 and 95 percent during the 4-day study):

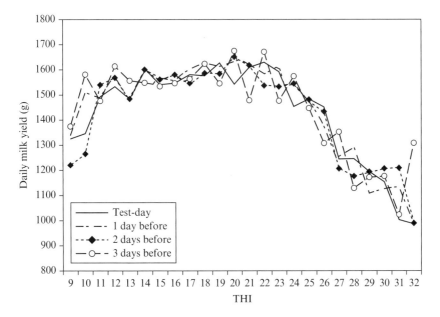

Source: Finocchiaro et al. (2005, p. 1860, Fig. 4).

Figure 2.8   *Relationship between the daily milk yield and the temperature*
          humidity index (THI) in various durations

$$THI = \{T - [0.55 \cdot (1 - RH)] \cdot (T - 14.4)\}.$$

Figure 2.8 presents the hill-shaped relationship between temperature–
humidity index and the daily milk yield. It can be seen that longer periods
of heat stress might have a more severe effect than shorter periods
(Finocchiaro et al., 2005). The study compared a 2, 3 and 4-day period of
heat stress. The study finds that the longevity of the heat stress period is
independent of yield. Milk yields were not significantly different across the
2, 3 and 4 days of heat stress.

There is a general recognition that energy demands increase in cold
environments and decrease in warm environments outside the optimal
range of EAT for each species. The optimal thermal environment for an
animal is associated with maximum performance and minimum stress and
disease. In a cold environment animals are able to invoke mechanisms
(such as changes in hair and feathers, and reduced blood flow in peripheral
vessels) to conserve body heat. However, this is reflected in less efficient
feed intake and lower performance. Usually, bigger producing animals

Table 2.2   Examples of thermoneutral ranges for various animal species

Species	Lower critical temperature (°C)	Higher critical temperature (°C)	Range (°C)
Hen	17	28	10
Hen (chick)	34	35	1
Sow	0	15	15
Sow (piglet)	32	33	1
Cow	0	16	16
Cow (calf)	12	24	12
Ewe	−3	20	23
Ewe (new born lamb)	29	30	1

Source: Based on NRC (1981, Figure 2).

with greater metabolic heat can endure larger differentials between body heat and cold environmental temperatures, and are therefore less susceptible to cold stress (NRC, 1981). In a warmer climate, animals face a reduced thermal gradient between their body and the environment, resulting in a reduced feed intake. Usually, bigger producing animals with larger metabolic heat have more trouble eliminating excess heat and are therefore more susceptible to heat stress (NRC, 1981). Note that baby animals tend to have very limited ranges of temperature within which they can survive (Table 2.2).

Generally speaking all species show a decreasing rate of feed intake as temperature rises. For example, dairy cattle intake rates decline from 140 to 60 percent of intake at 18–20°C (the benchmark within each species' thermoneutral zone) when temperature increases from −15°C to 40°C. Feedlot (beef) cattle intake rates decline from 105–115 to 80–50 percent of intake at 18–20°C when temperature increases from −18°C to 40°C. Poultry (laying hen) intake rates decline from 130 to 60 percent of intake at 18–25°C when temperature increases from −15°C to 35°C. The decline is not linear and declines more rapidly toward the higher range of temperature: >30°C for feedlot (beef) cattle, >25°C for dairy cattle, and >25°C for laying hen (NRC, 1981, pp. 27–34). Projected increases in both higher maximal temperatures and the incidences of extreme temperature events (IPCC, 2007) are likely to have an adverse affect on feeding rates.

At least for animals raised outdoors, climate also has indirect effects on animals by affecting the ecosystem. With different temperatures and rainfall, ecosystems will shift from forests, to savannahs, to grasslands to desert. These different ecosystems provide more carrying capacity for some animals compared to others. Climate will also affect insects and disease vectors, and it is likely that new disease vectors will be introduced into regions that have not previously experienced them. Places that are both hot and wet tend to be most problematic for livestock.

ROLE OF WATER

Just as it is for humans and plants, water is an important input for all animals. Some animals are given water directly. Animals also obtain water from dry feed, forage-pasture, rainfall, streams, dew and snow. Water intake is affected by feed intake and feed intake is affected by water intake. Water intake varies by species, breed, air temperature, water availability, water quality, water temperature, and other factors (NRC, 1981).

Climate Aspects Related to Dairy Cattle

In developed countries, commercial dairy cattle are held in cowsheds and are served the feed they need. Shelter can serve as a buffer for environmental extremes but, in turn, adds to energy costs. The amount of dry matter consumed by a milking cow depends on many variables, some of which can be controlled. These variables include health, live weight, milk production level, lactation stage, environmental conditions, management, and type of feed (NRC, 2001). In controlled settings, dairy cattle can be protected from high temperatures by cooling and by spraying.

Availability of water is also essential for dairy cattle health and productivity. Restricting the amount of water dairy cattle drink will restrict the intake of dry matter. On the other hand, the amount of water dairy cattle will consume is affected by the amount of dry matter ingested, and also by climate conditions. Water consumption also increases as air temperature rises and humidity level lowers. This relationship between humidity and temperature, and water intake reflects in part the cow's low intake of feed and in part the reduced vaporization of moisture at higher levels of humidity (NRC, 2001).

In developing countries, dairy cattle are often kept outdoors. They are therefore subject to the elements mentioned above, and climate will affect them directly. Climate can also affect them indirectly by affecting the ecosystem from which they must find food. Climate will cause ecosystems to shift between deserts, grasslands and forests, affecting the availability of food and water. Finally, disease vectors are climate-sensitive. Therefore, some hot and moist zones are more hostile to cattle.

Aspects Related to Beef Cattle

Beef cattle are raised mainly in the open although there are feedlots in developed countries. When the animals are in the open, they are directly affected by natural climates (NRC, 1981). Some of the non-thermal climatic conditions in the open have direct impact on beef cattle feed intake, and may lead to stress. For example, rain may lead to a temporary decline in feed intake of 10–30 percent; mud could lead to a decline in feed intake of 5–30 percent, depending on the mud depth.

Water is an important input dependent on various factors that include body weight, feed intake level, growth stage and air temperature. Air temperature has a marked impact on the level of water intake, ranging from 2–3 kg of water per 1 kg of dry matter feed intake at –5°C to 8–15 kg of water per 1 kg of dry matter feed intake (NRC, 1981, p. 62).

Aspects Related to Sheep

Climatic conditions such as temperature, humidity and wind may affect levels of energy needs of sheep, depending on other affecting conditions and on where in the thermoneutral zone the farm is located (NRC, 1985).

Most sheep production is on the range rather than in feedlots. Therefore, the exact relationship of feed and water intake and health and production is not known (NRC, 1985). Sheep get their water by drinking, from snow, dew and gustation, through metabolic water in feed or pasture. The amount of water consumed depends on many factors, including breed, production stage, temperature, rainfall, humidity and type of feed and pasture. Sheep are less sensitive to water supply compared to cattle. 'Nonbreeding Marino sheep have been maintained for two years without drinking water in a temperate climate with no reduction in wool production and body weight' (NRC, 1985, p. 19).

Range and pasture sheep husbandry is more difficult as both animals and plants have to be managed by the farmer. Moreover, climate affects the performance of both plants and animals, and the interaction among them. The precise relationship between climate and feed or pasture intake and health and productivity of range sheep is not known. This is an even bigger issue in arid and semi-arid regions where precipitation is associated with a more stochastic pattern.

Aspects Related to Poultry

Poultry, which includes hens, broilers, turkeys, geese, ducks and other species, are a relatively cheap source of protein for humans.[6] Climatic conditions, especially temperature, have significant impacts on energy requirements, and therefore on feed intake. Poultry feed intake is lower in hotter environments than in cold environments. Energy requirements for laying hens may vary substantially at different air temperature ranges (NRC, 1994). However, if temperatures become too high, chickens can no longer survive. Because of the shorter production cycle of poultry compared to other animals, dietary feed is adjusted to summer and winter climates. For example, broilers' production cycle is 3–6 months compared with 1–2 years for sheep and cattle.

While precise water needs for poultry are not known, water is an essential input. Poultry are sensitive to water intake and restrictions of water may lead to death, especially in younger animals (NRC, 1994).

CONCLUSION

These natural science analyses make clear that there is substantial evidence that greenhouse gases and climate affect both crops and animals. The impacts can be direct or indirect. The exact magnitude of the impacts appears to depend on many local conditions including soils, nutrients and water. Studies have also been unable to include all indirect effects including changes in ecosystems, weeds, insects and diseases. Therefore the scientific literature has not been able to provide conclusive evidence of the size of expected impacts in every location. However, there are some general results that are worth repeating. The relationship between yields and temperature appears to be hill-shaped for every species, though the 'best' temperature for each species is different. The steepness of this hill can be moderated by management adaptations. Agriculture in developing countries is likely to be more vulnerable than agriculture in developed countries because the climate is warmer and there are more limiting factors on production in developing country farms.

NOTES

1. This chapter builds on NRC (1981), Wolfe and Erickson (1994), Rotter and Van de Geijn (1999) and Reddy and Hodges (2000) for basic concepts associated with climate–animals and climate–crops interactions.
2. In some plants the roots (carrot) or the canopy (alfalfa) are considered 'yield'.
3. Several of these efficiency coefficients include (a) radiation efficiency measured in grams of dry weight per unit of solar radiation (megajoule); (b) water use efficiency measured in grams of dry weight per kilogram of water transpired; and (c) harvest efficiency measured in grams of yield per grams of total dry matter.

4. Rubisco enzyme activity is reduced in crops exposed to long-term high CO_2 concentration and low temperature (Wolfe and Erickson, 1994).
5. There have been several recent studies criticizing Long's analysis. It was argued that in comparing the FACE system to all other methodologies one finds that, with the exception of single plant studies, the FACE results are skewed. Additionally, it was argued that Long did not account for varietal variability, or pests, diseases, weeds, and so on.
6. It takes 20 pounds of grain to produce one pound of beef, 7 pounds of grain for one pound of pork, 4 pounds of grain for one pound of chicken, and 1.2 pounds of grain for one pound of fish. So, fish may be the cheapest source of protein, if one does not take into consideration the investment in necessary infrastructure.

REFERENCES

Allen, L.H. and K.J. Boote (2008), personal communication, based in part on data published in Baker et al. (1992a, b).

Baker, J.T., L.H. Allen and K.J. Boote (1992a), 'Effects of CO_2 and temperature on growth and yield of rice', *Journal of Experimental Botany*, **43**, 959–64.

Baker, J.T., L.H. Allen and K.J. Boote (1992b), 'Response of rice to carbon dioxide and temperature', *Agricultural and Forest Meteorology*, **60**, 153–66.

Bunce, J.A. and L.H. Ziska (2000), 'Crop ecosystem responses to climatic change: crop/weed interactions', in K.R. Reddy and H.F. Hodges (eds), *Climate Change and Global Crop Productivity*, Wallingford, UK: CABI Press, pp. 333–52.

Cheikh, N., P.W. Miller and G. Kishore (2000), 'Role of biotechnology in crop productivity in a changing environment', in K.R. Reddy and H.F. Hodges (eds), *Climate Change and Global Crop Productivity*, Wallingford, UK: CABI Press, pp. 425–36.

Colhoun, J. (1973), 'Effects of environmental factors on plant disease', *Annual Review of Phytopathology*, **11**, 343–64.

Cure, J.D., D.W. Israel and T.W. Rufty, Jr. (1988), 'Nitrogen stress effects on growth and seed yield of non-nodulated soybean exposed to elevated carbon dioxide', *Crop Science*, **28**, 671–77.

Darwin, R. and D. Kennedy (2000), 'Economic effects of CO_2 fertilization of crops: transforming changes in yield into changes in supply', *Environmental Modeling and Assessment*, **5**(3), 157–68.

Finocchiaro, R., J.B.C.H.M. van Kaam, B. Portolano and I. Misztal (2005), 'Effect of heat stress on production of Mediterranean dairy sheep', *Journal of Dairy Science*, **88**, 1855–64.

Fuhrer, J. (2003), 'Agro-ecosystem responses to combinations of elevated CO_2, ozone and global climate change', *Agriculture, Ecosystems and Environment*, **97**, 1–20.

Grace, John and Rui Zhang (2006), 'Predicting the effects of climate change on global plant productivity and the carbon cycle', in James I.L. Morison and Michael D. Morecroft (eds), *Plant Growth and Climate Change*, Oxford, UK: Blackwell Publishing, pp. 187–208.

Gutierrez, A.P. (2000), 'Crop ecosystem responses to climatic change: pests and population dynamics', in K.R. Reddy and H.F. Hodges (eds), *Climate Change and Global Crop Productivity*, Wallingford, UK: CABI Press, pp. 353–74.

Hall, A.E. and L.H. Ziska (2000), 'Crop breeding strategies for the 21st century', in K.R. Reddy and H.F. Hodges (eds), *Climate Change and Global Crop Productivity*, Wallingford, UK: CABI Press, pp. 407–19.

Hartwell, Allen Jr. (2008), personal communication, 19 September.

Hartwell, Allen Jr. and K.J. Boote (2000), 'Crop ecosystem responses to climatic change: soybean', in K.R. Reddy and H.F. Hodges (eds), *Climate Change and Global Crop Productivity*, Wallingford, UK: CABI Press, pp. 133–60.

Horie, T., J.T. Baker, F. Nakagawa, T. Matsuui and H. Y. Kim (2000), 'Crop ecosystem responses to climatic change: rice', in K.R. Reddy and H.F. Hodges (eds), *Climate Change and Global Crop Productivity*, Wallingford, UK: CABI Press, pp. 81–106.

IPCC (2007), *Climate Change 2007: The Physical Science Basis*, contribution of Working Group I to the Fourth Assessment Report of the Intergovernmental Panel on Climate Change, Cambridge, UK: Cambridge University Press.

Janssens, I.A., M. Mousseau and R. Ceulemans (2000), 'Crop ecosystem responses to climatic change: tree crops', in K.R. Reddy and H.F. Hodges (eds), *Climate Change and Global Crop Productivity*, Wallingford, UK: CABI Press, pp. 245–70.

Jones, Gerald M. and Charles C. Stallings (1999), 'Reducing heat stress for dairy cattle', Virginia Cooperative Extension, Publication 404-2000, www.ext.vt.edu/pubs/dairy/404-200/404-200.html, accessed 8 February, 2008.

Ke, B. (2001), *Photosynthesis Photobiochemistry and Photobiophysics: Advances in Photosynthesis and Respiration*, Vol. 10, Berlin: Springer.

Kim, H.Y., T. Horie, H. Nakagawa and K. Wada (1996), 'Effects of elevated CO_2 concentration and high temperature on growth and yield of rice', *Japanese Journal of Crop Science*, **65**, 644–51 (Japanese with abstract in English).

Kim, H.Y., M. Lieffering, K. Kobayashi, M. Okada and S. Miura (2003), 'Seasonal changes in the effects of elevated CO_2 on rice at three levels of nitrogen supply: a free air CO_2 enrichment (FACE) experiment', *Global Change Biology*, **9**, 826–37.

Kimball, B.A. (1983), 'Carbon dioxide and agricultural yields: an assemblage and assessment of 430 prior observations', *Agronomy Journal*, **75**, 779–88.

Kimball, B.A. (2007), 'Review of "Plant growth and climate change"', *Quarterly Review of Biology*, **82**(4), 436–37.

Krishnan, P., D.K. Swain, H. Chandra Bhaskar, S.K. Nayak and R.N. Dash (2007), 'Impact of elevated CO_2 and temperature on rice yield and methods of adaptation by crop simulation studies', *Agriculture Ecosystems & Environment*, **122**, 233–42.

Lawlor, T.W. and R.A.C. Mitchell (2000), 'Crop ecosystem responses to climatic change: wheat', in K.R. Reddy and H.F. Hodges (eds), *Climate Change and Global Crop Productivity*, Wallingford, UK: CABI Press, pp. 57–80.

Long, S.P., E.A. Ainsworth, A.D.B. Leakey and P.B. Morgan (2005), 'Global food insecurity. Treatment of major food crops with elevated carbon dioxide or ozone under large-scale fully open-air conditions suggests recent models may have overestimated future yields', *Philosophical Transactions of the Royal Society of London. Series B, Biological Sciences*, **360**, 2011–20.

Long, S.P., E.A. Ainsworth, A.D.B. Leakey, J. Nösberger and D.R. Ort (2006), 'Food for thought: lower-than-expected crop yield stimulation with rising CO_2 concentrations', *Science*, **312**, 30 June, 1918–21.

Manion, P.D. and D. LaChance (1992), *Forest Decline Concepts*, St. Paul, MN, USA: American Phytopathological Society Press.

Matsui, T., O.S. Namuco, L.H. Ziska and T. Horie (1997), 'Effects of high temperature and CO_2 concentration on spikelet sterility in Indica rice', *Field Crops Research*, **51**, 213–21.

Miglietta, F., M. Bindi, F.P. Vaccari, A.H.C.M. Schapendonk, J. Wolf and R.E. Butterfield (2000), 'Crop ecosystem responses to climatic change: root and tuberous crops', in K.R. Reddy and H.F. Hodges (eds), *Climate Change and Global Crop Productivity*, Wallingford, UK: CABI Press, pp. 189–212.

Morison, James I.L. and Michael D. Morecroft (2006), *Plant Growth and Climate Change*, Oxford, UK: Blackwell Publishing.

Morison, James I.L. (1996), 'Climate change and crop growth', *Environmental Management and Health*, **7**(2), 24–7.

National Research Council (NRC), Subcommittee on Beef Cattle Nutrition, Committee on Animal Nutrition Board on Agriculture (2000), *Nutrient Requirements of Beef Cattle*, (7th revised edn), Washington, DC, USA: National Academic Press.

National Research Council (NRC), Subcommittee on Dairy Cattle Nutrition, Committee on Animal Nutrition Board on Agriculture (2001), *Nutrient Requirements of Dairy Cattle*, (7th revised edn), Washington, DC, USA: National Academic Press.

National Research Council (NRC), Subcommittee on Environmental Stress (1981), *Effect of Environment on Nutrient Requirements of Domestic Animals*, Washington, DC, USA: National Academic Press.

National Research Council (NRC), Subcommittee on Poultry Nutrition, Committee on Animal Nutrition Board on Agriculture (1994), *Nutrient Requirements of Poultry*, (9th revised edn), Washington, DC, USA: National Academic Press.

National Research Council (NRC), Subcommittee on Sheep Nutrition, Committee on Animal Nutrition Board on Agriculture (1985), *Nutrient Requirements of Sheep*, (6th revised edn), Washington, DC, USA: National Academic Press.

Newton, P.C.D. and G.R. Edwards (2007), 'Plant breeding for a changing environment', in P.C.D. Newton, R.A. Carran, G.R. Edwards and P.A. Niklaus (eds), *Agroecosystems in a Changing Climate*, Boca Raton, FL, USA: CRC Press, 309–22.

Nosberger, J., H. Blum and J. Fuhrer (2000), 'Crop ecosystem responses to climatic change: productive grasslands', in K.R. Reddy and H.F. Hodges (eds), *Climate Change and Global Crop Productivity*, Wallingford, UK: CABI Press, pp. 271–91.

Parry, M.L., J.H. Porter and T.R. Carter (1990), 'Agriculture: climatic change and its implications', *Trends in Ecology and Evolution*, **5**, 318–22.

Patterson, D.T., J.K. Westbrook, R.J.C. Joyce, P.D. Lingren and J. Rogasik (1999), 'Weeds, insects and diseases', *Climatic Change*, **43**, 711–27.

Peet, M.M. and D.W. Wolf (2000), 'Crop ecosystem responses to climatic change: vegetable crops', in K.R. Reddy and H.F. Hodges (eds), *Climate Change and Global Crop Productivity*, Wallingford, UK: CABI Press, pp. 213–44.

Pereira, João S., Maria-Manuela Chaves, Maria-Conceição Caldiera and Alexander V. Correira (2006), 'Water availability and productivity', in James I.L. Morison and Michael D. Morecroft, *Plant Growth and Climate Change*, Oxford, UK: Blackwell Publishing, pp. 118–45.

Polley, H.W. (2002), 'Implications of atmospheric and climatic change for crop yield and water use efficiency', *Crop Science*, **42**, 131–40.

Prasad, P.V.V., K.J. Boote and L.H. Allen Jr. (2006), 'Adverse high temperature effects on pollen viability, seed-set, seed yield and harvest index of grain-sorghum [Sorghum bicolor (L.) Moench] are more severe at elevated carbon dioxide due to high tissue temperature', *Agricultural and Forest Meteorology*, **139**, 237–51.

Prasad, P.V.V., K.J. Boote, L.H. Allen Jr. and J.M.G. Thomas (2002), 'Effects of elevated temperature and carbon dioxide on seed-set and yield of kidney bean (*Phaseolus vulgaris* L.), *Global Change Biology*, **8**, 710–21.

Reddy, K.R. and H.F. Hodges (eds) (2000), *Climate Change and Global Crop Productivity*, Wallingford, UK: CABI Press.

Reddy, K.R., H.F. Hodges and B.A. Kimball (2000), 'Crop ecosystem responses to climatic change: cotton', in K.R. Reddy and H.F. Hodges (eds), *Climate Change and Global Crop Productivity*, Wallingford, UK: CABI Press, pp. 161–88.

Rosenzweig, C. and D. Hillel (1995), *Climate Change and the Global Harvest: Potential Impacts on the Greenhouse Effect on Agriculture*, New York, USA: Oxford University Press.

Rotter, R., S.C. van de Geijn (1999), 'Climate change effects on plant growth, crop yield and livestock', *Climatic Change*, **43**(4), 651–81.

Royle, D.J., M.W. Shaw and R.J. Cook (1986), 'The natural development of *Septoria nodorurn* and *S. tritici* in some winter wheat crops in Western Europe, 1981–83', *Plant Pathology*, **35**, 466–76.

Treharne, K. (1989), 'The implications of the "greenhouse effect" for fertilizers and agrochemicals', in R.C. Bennet (ed.), *The Greenhouse Effect and UK Agriculture*, London, UK: Ministry of Agriculture, Fisheries and Food, UK, pp. 67–78.

United States Department of Agriculture (USDA) and US Climate Change Science Program (USCCSP) (2008), *The Effects of Climate Change on Agriculture, Land Resources, Water Resources, and Biodiversity*, Washington, DC: USDA–ARS.

Wahaj, R., M. Florent and M. Giovanni (2007), 'Actual crop water use in project countries: a synthesis at the regional level', World Bank Policy Research Working Paper 4288, Washington, DC, USA: World Bank.

Wolf, J. (1996), 'Effects of nutrient supply (NPK) on spring wheat response to elevated atmospheric CO_2', *Plant and Soil*, **185**, 113–23.

Wolfe, David W. and Jon D. Erickson (1994), 'Carbon dioxide effects on plants', in Harry M. Kaiser and Thomas E. Drennen (eds), *Agricultural Dimensions of Global Climate Change*, Delray Beach, FL, USA: St. Lucie Press.

Young, K.G. and S.P. Long (2000), 'Crop ecosystem responses to climatic change: maize and sorghum', in K.R. Reddy and H.F. Hodges (eds), *Climate Change and Global Crop Productivity*, Wallingford, UK: CABI Press, pp. 107–32.

Ziska, L.H. and J.A. Bunce (2006), 'Plant responses to rising atmospheric carbon dioxide', in J.I.L. Morison and M.D. Morecroft (eds), *Plant Growth and Climate Change*, Oxford, UK: Blackwell Publishing, pp. 17–47.

Ziska, L.H. and G.B. Runion (2007), 'Future weed, pest, and disease problems for plants', in P.C.D. Newton, R.A. Carran, G.R. Edwards and P.A. Niklaus (eds), *Agroecosystems in a Changing Climate*, Boca Raton, FL, USA: CRC Press, pp. 261–85.

Ziska, L.H. and J.R. Teasdale (2000), 'Sustained growth and increased tolerance to glyphosate observed in a C3 perennial weed, quackgrass (Elytrigia repens), grown at elevated carbon dioxide', *Australian Journal of Plant Physiology*, **27**, 159–64.

Ziska, L.H., J.R. Teasdale and J.A. Bunce (1999), 'Future atmospheric carbon dioxide may increase tolerance to glyphosate', *Weed Science*, **47**, 608–15.

3. Literature review of economic impacts of climate change on agriculture[1]

In the previous chapter we reviewed the scientific literature that links climate to both crop and livestock productivity. This chapter reviews the literature that measures the economic impact of climate change on agriculture. Many studies of economic impacts of climate change are concerned that changes in climate will have harmful effects on the agricultural sector (for example Smith and Tirpak, 1989; Nordhaus, 1991; Cline, 1992; Rosenzweig and Parry, 1994; Pearce et al., 1996; Tol, 2002; Mendelsohn and Williams, 2004; Parry et al., 2004). Some of these studies predict there will be adverse agricultural impacts in all countries, whereas others expect the harmful impacts to be concentrated in developing countries. Developing countries may be more vulnerable to climate change because these regions are already very hot and the agricultural systems may be less capable of adapting.

We distinguish amongst six main types of approaches: (1) studies that rely on crop simulation models; (2) studies that rely on cross-sectional or intertemporal analyses of yields; (3) agro-economic simulation models of farms; (4) studies based on panel (intertemporal) analysis of net revenues across weather; (5) studies based on cross-sectional analyses of net revenues or land values per hectare; and (6) studies that use CGE (computable general equilibrium) models. In the following sections, we discuss each of these approaches in detail and highlight their strengths and weaknesses.

CROP SIMULATION MODELS

The first and one of the most popular methods for estimating the impacts of climate change on agriculture relies on crop simulation models. Crop simulation models use functions that capture the interaction between crop growth and climate, soils and management practices. The crop simulation models are calibrated to selected locations. Different climate change scenarios are run for each location for selected crops given a particular management practice. The yield changes are then extrapolated to an aggregate effect (Rosenzweig et al., 1993; Rosenzweig and Parry, 1994; Parry et al., 1994; Rosenzweig and Iglesias, 1998). Below we review the major brands of models (based on the review by Iglesias and Garrote, 2007):

The CERES-Maize simulation model is a model of maize (for example, Mearns et al., 1996, 1999; Kapetanaki and Rosenzweig, 1997; Ritchie et al., 1998). The model predicts the response of the corn plant to different soils and weather/climate given the management conditions. The model is well suited to addressing the impacts of weather/climate variability (Mavromatis and Jones, 1998; Mearns et al., 1999), and can be used to make management decisions (Jones et al., 2000; Hansen et al., 2001).

The ICASA/IBSNAT[2] dynamic crop growth models are decision support systems that simulate crop responses to management practices (DSSAT). Such models are particularly useful for evaluating adaptation options to climate change. The ICASA/IBSNAT models have been used widely for evaluating climate impacts in agriculture at different levels ranging from individual sites to wide geographic areas (Rosenzweig and Iglesias, 1998). The DSSAT software includes all ICASA/IBSNAT models with an interface that allows output analysis.

The EPIC model (Erosion Productivity Impact Calculator), developed by Sharpley and Williams (1990), is a model that was developed for evaluation of management practices in conjunction with soil qualities. It incorporates simplified crop growth functions that respond to climate, environment and management.

CROPWAT is an empirical irrigation management model developed by the United Nations Food and Agriculture Organization (FAO) to calculate regional crop water and irrigation requirements based on climatic data and crop parameters (CROPWAT, 1995, 2004). The advantage of this model is its simplicity and ease of application, using the FAOCLIM global database (FAO, 2004). The model can be adjusted to include irrigation efficiency for each region, simulating technological adaptation. In a recent upgrade, the CROPWAT model was extended (Wahaj et al., 2007) to include impact of CO_2 on yields, water and season length.

Quite a large number of applications of the above models have been reported in the literature (Iglesias and Garrote, 2007). For example, yield responses to climate and management are simulated at selected sites using DSSAT (Rosenzweig and Iglesias, 1998). The impact of climate change and CO_2 concentration on crops and water was estimated for Spain (Guereña et al., 2001). Iglesias and Garrote (2007) quantify the impact of climate change on selected crops in several European agricultural regions. A comprehensive set of various agronomic studies by country can be found at www.climate.org/CI/ag/regional.shtml.

The crop simulation models are attractive because they are developed based on a deep understanding of agronomic science. They can be carefully linked with hydrologic conditions. They can integrate the effects of carbon dioxide fertilization. Finally, they are calibrated to local conditions. There has been particularly thorough examination of the impacts of climate change on the major grains (Rosenzweig and Parry, 1994; Rosenzweig and Iglesias, 1998) including maize (Muchena, 1994; Muchena and Iglesias, 1995; Mearns et al., 1996, 1999; Kapetanaki and Rosenzweig, 1997; Ritchie et al., 1998), wheat (Barrow and Semenev, 1995; Mearns et al., 1996, 1999), soybeans (Lal et al., 1999), and rice (Jin et al., 1994; Horie et al., 2000; Saseendraen et al., 2000).

However, crop simulation models also have important limitations. One of the most significant of these concerns adaptation. The underlying crop simulation model is purely an agronomic relationship. The model does not capture the behavior of the farmer; the management practice of the farmer is assumed to be exogenous or fixed. If farmers continue to behave as they did when the model was calibrated, the results are accurate. However, the model does not predict how farmers are likely to change their behavior as climate changes. To the extent that adaptation is included in crop simulation studies, it is introduced exogenously by the researcher. For example, the most comprehensive crop simulation studies have assumed that all farmers might add fertilizer to partially compensate for the loss in yields (Rosenzweig and Parry, 1994; Parry et al., 2004, 2005). The problem with these examples of adaptation is that they are arbitrary. They are neither targeted at climate change nor are they motivated by profit maximization.

This limitation in the crop simulation literature can be overcome. Several studies using crop simulation models have examined possible responses by the farmer to climate change. For example, Jin et al. (1994) find that using new rice cultivars and changing planting dates in Southern China can substantially increase rice yields. You (2001) finds that switching from rice to corn in China has the potential to save 7 billion m^3 of water per year in China. Changing planting dates and varieties of maize can increase yields by 10 percent in Greece (Kapetanaki and Rosenzweig, 1997). Hybrid seeds and altered sowing dates can allow for double cropping for wheat and maize in Spain, increasing yields and reducing water use (Iglesias and Minguez, 1996). Stuczyinski et al. (2000) conclude that Polish agriculture production could avoid declines of 5–22 percent of production by engaging in adaptation strategies. Mizina et al. (1997) find adaptation could help Kazakhstan avoid 70 percent reductions in yields from climate change. Note that these adaptations are designed for local conditions. However, even these examples are not motivated by profit maximization.

A second limitation of the crop simulation models is that they model one crop at a time. The crop simulation models have not been used to predict how farmers would react as one crop becomes less suitable and another crop becomes more suitable. They do not predict crop switching, despite its clear importance in climate change.

A third limitation is that only a limited number of crops have been studied. Researchers have targeted the major grains because they are the largest single crops. However, the major grains tend to be grown in cool or temperate climates (with the exception of irrigated rice). Crops that are more suited to warm climates such as fruits and vegetables have been omitted. The studies are consequently biased towards finding that warming is harmful.

Finally, crop simulation models have been calibrated only in a limited number of places. If these locations are not representative of all farms, they can provide misleading predictions.

EMPIRICAL YIELD MODELS

Another way to measure the sensitivity of yields to climate is to measure how actual yields vary under different climate conditions. For example, one can conduct cross-sectional studies of actual yields across different climate zones. By applying an empirical production function model, one can isolate the effects of climate from other factors influencing yields. The production function approach links water, soil, climate and economic inputs, to crop yields for specific crops. For example, Onyeji and Fischer (1994) consider the impacts of climate change on yields in Egypt with and without adaptation. Gbetibouo and Hassan (2005) apply a cross-sectional approach to estimate the climate sensitivity for seven field crops in South Africa. Econometric methods have also been used to predict the climate sensitivity of yields of sorghum and corn in the USA (Sands and Edmonds, 2005; Chavas et al., 2001). The US studies find that future climate change would have a positive effect on sorghum yields but a negative effect on corn yields.

Another way to empirically measure the sensitivity of yields is to examine the effect of weather on yields over time. The first study to do this relied on a unique weather condition in the middle of the USA in the 1930s called the 'dustbowl' (Crosson, 1993; Easterling et al., 1993). For a brief period, temperatures were higher and precipitation slightly lower, leading to unusually dry soil conditions in this region. The study measures the reduction in yields of selected grains in this period compared to periods with normal weather across the region.

The empirical yield function approach, however, has some of the same limitations as the crop simulation approach. Farmers are often assumed to continue growing the same crop, with the same technology regardless of the change in climate. The analyses often focus only on a limited set of crops. The full set of adaptations available to farmers is underestimated.

ECONOMIC MANAGEMENT MODELS

In order to capture farm behavior, it is important to model the farm and the farming sector, not just the plants. Farm simulation models assume that farmers wish to maximize their profits. One can then determine what farm behavior would lead to profit maximization. One can also determine which farm adaptations would maximize profits in response to climate change. For example, farmers might alter planting times, crop varieties, harvest dates, tillage and irrigation methods. The researcher would then be able to determine which activity would maximize profit and then could trace out actual yields and net revenues for different climates. In practice, the farm simulation models find that looking at all alternatives is too expensive, and so they tend to include only a limited number of alternative farming methods. The farm simulation models thus begin to address the critical issue of adaptation. However, it is likely that the models do not yet fully capture the full range of adaptations that farmers can practice.

Farm simulation models can capture the behavior of a single farm (Kaiser et al., 1993) or all farms in a country (Adams et al., 1990; 1993; 1999). The single farm model can describe in detail the alternative choices a farmer might make to maximize profits at a specific location. A national farm sector model can describe how these farm choices vary from place to place and also how farm prices are likely to change (although changes in farm prices require assumptions about changes in other countries).

In practice, almost all farm simulation models have focused on the USA. In this review we focus on just the models that have been used to study climate change. The farm simulation models tend to rely on crop simulation models to make the link between changes in climate and CO_2 and crop yields. Because this link may also be a function of other management choices by farmers, such as when to plant, the models are not able to capture the full range of adaptation available. Mathematical programming methods are used to determine profit maximizing choices of crops by farmers across the USA (Adams et al., 1990, 1993, 1995, 1999) or on a single farm (Kaiser et al., 1993; Mount and Li, 1994). These studies predict crop switching as prices or climate change.

Another group of studies relying on the farm economic model have been conducted under the leadership of John Reilly for the US Global Change Research Program (Reilly et al., 2001). The studies conclude that by the end of the century, with climate change and adaptation, US rainfed agricultural yields will rise on average by 89 percent for cotton, 29 percent for corn, 80 percent for soybeans, 24 percent for wheat, and 11 percent for rice, and would decline for potatoes by 11 percent. Irrigated yields with adaptation were estimated to rise by 110 percent for cotton, 4 percent for corn, 36 percent for soybeans, 4 percent for wheat, and 11 percent for rice, but to decline by 14 percent for potatoes.

A study of Taiwan provides a rare example of a farm economic model outside the US (Chang, 2002). Yield response regressions for 60 crops are incorporated into a farm model with climate and environmental characteristics of various regions in Taiwan. The model is then used to examine climate change. Although Taiwan is a relatively small island, the impact of climate change on agriculture is shown to vary significantly by regions.

INTERTEMPORAL NET REVENUE APPROACH

Instead of empirically measuring how yields change over time, one can instead focus on measuring the changes in net revenue. Using panel data, one can estimate the effect of changes in weather on farm net revenue (Deschenes and Greenstone, 2007). By using fixed effects to control for the differences between one county and another, this approach controls for all the permanent differences between counties, including climate, as well as other differences that are hard to measure. However, at the same time, fixed effects also control for adaptations that farmers have made to adjust to climate. The analysis focuses only on the intertemporal differences between counties which reflect year to year variations in weather. This approach in principle is an ideal method to measure short-term responses to sudden changes in weather. However, it suffers from all the limitations of the yield studies in that it does not reflect adaptation. The empirical example (in Deschenes and Greenstone, 2007) is further hampered by the inclusion of state time dummies that remove most of the observed variation in weather facing each farmer.

RICARDIAN (CROSS-SECTION) ANALYSES

The Ricardian technique (for a detailed discussion see Chapter 4), named after David Ricardo (1772–1823), estimates the net productivity of farmland as a function of climate, soils and other control variables (Mendelsohn et al., 1994). The technique relies on a cross-sectional sample of farms that span a range of climates, to measure the sensitivity of land value or farm net revenue per hectare to climate.

With the Ricardian technique, land value or net revenue is regressed on a set of climate variables (for example, rainfall and temperature measured either in annual or seasonal values), other environmental characteristics (for example, soil and altitude), exogenous market factors, and other control variables. By controlling properly for unwanted correlation, the regression can measure the impact of the climatic variables. Note that farm choices such as labor, capital and crop choice are not included in the Ricardian regression because they are endogenous. As with all empirical methods, the more accurate the measurements of the variables, the better uncontrolled variables are accounted for, the more variation in the desired variables (climate), and the larger the sample, the more accurate the results.

The Ricardian method was first applied to study land values in the United States (Mendelsohn et al., 1994). There has been a long series of additional analyses of the US using more recent observations and additional data. For example, Mendelsohn et al. (1996) examine whether the quantity of cropland (not just the value per hectare) is also sensitive to climate. Mendelsohn et al. (1999) examine the role of climate variance. Mendelsohn and Dinar (2003) examine the role of water withdrawals. Schlenker et al. (2005) examine whether counties with rainfed farms have similar responses to all counties. They find that the counties with rainfed farms have greater sensitivity to higher temperatures. Schlenker et al. (2006, 2007) use degree days during the growing season instead of seasonal temperatures. They find that degree days explain a great deal of the variation amongst counties. However, they do not address the fact that degree days are endogenous measurements that depend upon the growing season, which will change as climate changes. Deschenes and Greenstone (2007) examine Ricardian analyses over time and note that the coefficients are not stable over time.

The earliest Ricardian studies in developing countries analyzed Brazil and India (Sanghi et al., 1997; Dinar et al., 1998; Kumar and Parikh, 2001; Sanghi et al., 1998; Mendelsohn and Dinar, 1999; Mendelsohn et al., 2001). These countries had the advantage of existing data sets and large territories with significant variation in climate.

The Ricardian method has also been applied to a number of other locations. The first Ricardian study in Africa was done by Ouedraogo (1999) in Morocco and later by Balti (2001) using data from Spain and Tunisia. There has also been a large study of 11 countries in Africa (Dinar et al., 2008). The African analysis focused on net revenue per hectare for individual farms. One set of these studies focused on crop net revenue. The studies produced Ricardian regressions for crop net revenue across 11 countries (Kurukulasuriya et al., 2006; Kurukulasuriya and Mendelsohn, 2008b) as well as a series of single country studies in Burkina Faso (Ouedraogo et al., 2007), Cameroon (Molua and Lambi, 2007), Ethiopia (Deressa, 2007), Egypt (Eid et al., 2007), Kenya (Kabubo-Mariara and Karanja, 2007), Senegal (Sene et al., 2007), South Africa (Benhin, 2007), Zambia (Jain, 2007) and Zimbabwe (Mano and Nhemachena, 2007).

The African analysis also examined the net revenue of livestock production across the 11 countries (Seo and Mendelsohn, 2008a). This study showed that livestock net income increased as temperatures fell and as rainfall fell. The livestock net revenue consequently is a balance against the crop net revenue protecting farmers from climate change.

The African data was also used to look at the importance of AgroEcological Zones (AEZs). The entire continent has been broken down into AEZs in order to assess agricultural potential (FAO, 1978). Seo et al. (2008a) explore how climate affects the income per farm in each AEZ. The climate sensitivity by AEZ is then used to extrapolate from the sample of farms studied to the entire continent. This methodology provides a detailed map of how climate is likely to affect agriculture across the continent.

There has been a seven-country Ricardian study of South America (Mendelsohn et al., 2007a). Ricardian regressions were run for all seven countries (Seo and Mendelsohn, 2008a) as well as for Argentina (Lozanoff and Cap, 2006), Brazil (Avila et al., 2006), Chile (González and Velasco, 2008), Columbia (Granados et al., 2006), Ecuador (Jativa and Seo, 2006), Uruguay (Lanfranco and Lozanoff, 2006), and Venezuela (Pacheco, 2006).

There have also been studies in other developing countries. Despite its small size, there have been two studies in Sri Lanka (Kurukulasuriya and Ajwad, 2007; Seo et al., 2005). There have also been two studies in China (Liu et al., 2004 and Wang et al., 2008).

There have been additional Ricardian studies in more developed countries. There have been studies of England and Wales (Maddison, 2000), Germany (Lang, 2001), Canada (Reinsborough, 2003; Weber and Hauer, 2003), and Israel (Fleischer et al., 2008). There have also been multi-country studies of Brazil and the US to study satellite data (Mendelsohn et al., 2007b); Brazil, India and the US were analyzed to study climate variance (Mendelsohn et al., 2007c); Brazil and the US were combined to study rural income (Mendelsohn et al., 2007d); and the US and Canada to study similarities (Mendelsohn and Reinsborough, 2007).

One of the principal advantages of the Ricardian technique is that it incorporates efficient adaptations by farmers to climate change. The approach does not suffer from the ad hoc adaptation adjustments of all the other approaches. The technique captures how farmers modify their production practices in response to changes in water availability, rainfall pattern and temperature. However, even this advantage must be treated with caution. The Ricardian method measures how farmers currently would adapt to climate. But the ability of farmers to adapt to climate is based on many factors that may change over time, including technology, input and output prices, human capital, infrastructure, and support services by the government. What is most important from a policy perspective is how future farmers will adapt to the future climate. The Ricardian technique does not necessarily predict future adaptation.

There have been many criticisms leveled at the Ricardian technique. The technique does not fully control for the impact of important variables that could also explain the variation in farm incomes, especially irrigation (Darwin, 1999; Schlenker et al., 2005). Subsequent studies have tried to address irrigation in a number of ways including measuring surface water (Mendelsohn and Dinar, 2003), modeling irrigation (Kurukulasuriya and Mendelsohn, 2008c), and separating rainfed from irrigated farms (Schlenker et al., 2005; Kurukulasuriya and Mendelsohn, 2008b; Seo and Mendelsohn, 2008b). A second criticism is that the Ricardian technique assumes prices are constant when prices may in fact change (Cline, 1996). This assumption leads to a small bias that causes the method to overestimate damages and benefits (Mendelsohn and Nordhaus, 1999b). A third critique is that the Ricardian method does not measure the adjustment costs from one equilibrium to another (Kaiser et al., 1993; Quiggin and Horowitz, 1999; Kelly et al., 2005). This is an important caveat to remember. The Ricardian method is a comparative steady state analysis of long-term climate impacts and not a dynamic analysis of short-term weather effects. Other approaches such as the intertemporal production function and net revenue function are better at measuring short-term weather effects. A fourth criticism is if land within observed locations is heterogeneous, migration may cause bias (Timmins, 2006).

Because this book focuses on the Ricardian method, we do not go into more detail about the method here. Both the Ricardian methodology and the empirical results are examined in much greater detail in other chapters. This chapter merely presents a brief review.

CGE MODELS

The five approaches reviewed so far are based on partial equilibrium models of the agriculture sector. However, if changes induced by climate change are large enough, they may affect the entire economy and change both input and output prices. An economy-wide approach may be needed to capture these broader changes. Computable General Equilibrium (CGE) models are used to capture economy-wide and global changes.

The clear advantage of the CGE models is that they can predict how large shifts in supply and demand can alter prices and thus capture changes that the partial equilibrium models miss. Although the partial equilibrium models might capture shifts within the sector such as in crop prices, they invariably miss changes in interest rates or labor prices. The partial equilibrium models also tend to be country-specific so that they miss trade effects as well.

One of the disadvantages of moving to global models is underlying inaccuracies. The models depend on accurate measures of the underlying sensitivity of each sector to climate change. The models also need to have reliable estimates of how the different sectors of the economy interact. These inaccuracies are compounded with aggregation. It is very difficult for CGE modelers to predict how climate affects supply in large regions, much less regions that are poorly studied, such as developing countries. With the bulk of climate damages occurring in developing countries, one must treat CGE results with caution. For example, Darwin et al. (1995, 1996) use a global model that includes eight world regions (US, Canada, European Community, Japan, China and several other East Asian Countries, some Southeast Asian countries, Australia and New Zealand, and the rest of the world). All the developing countries are consequently treated as a single region in this model. The CGE model aggregates information on land and climatic resource changes based on a Geographic Information System (GIS) and changes in climate that are predicted by GCMs. Although comprehensive, the CGE model requires detailed knowledge of land and agricultural uses which cannot be accurately measured at this level of aggregation. Further, by dealing with large land areas as units affected by climate change, the results mask important climate–agriculture relationships.

More regionally-based studies can help address these shortcomings by verifying assumptions with more detailed country-level analysis. Winters et al. (1998) studied the impact of global climate change on developing countries using CGE-multimarket models for three archetypal economies representing the poor cereal-importing nations of Africa, Asia and Latin America. The objective was to compare the effects of climate change on macroeconomic performance, sectoral resource allocation, and household welfare across continents. The results suggest that all these developing countries are likely to suffer income and production losses. Africa, with its low substitution possibilities between imported and domestic foods, is predicted to have the largest income losses and the largest drop in consumption amongst low-income households. Policy interventions that were found effective include integration in the international market, improved production of food crops in Africa, and improved production of export crops in Latin America and Asia.

Jorgensen et al. (2004) apply a CGE framework to the US economy. This analysis builds on an earlier effort (Scheraga et al., 1993) that estimated the aggregate economic effects of climate change. Assumed temperature and sea level relationships were used to generate ranges, rates and levels of damages (or benefits) associated with alternative climate change scenarios. The Jorgenson et al. (2004) model includes additional features such as climate impacts on water supply and human health. Key market impacts associated with these scenarios and damage relationships were simulated using a detailed model of the US economy known as the Inter-temporal General Equilibrium Model (IGEM). IGEM integrates the changes predicted for each sector into an aggregate response by the entire economy. Of course, it cannot be any more accurate than the assumed relationships between climate and each sector.

CONCLUSION

The economic assessment of the impacts of climate change on agriculture is a very difficult task. The impacts are likely to vary a great deal across space and across time. The effects are likely to depend not only on climate but also on other local characteristics such as soil, surface water, technology and market access. No matter what methodology is used, there are many assumptions that have to be made, some of which may affect the nature of the findings. There is no single approach that dominates all others. Many approaches have been developed in the literature and they all provide some needed insight.

Although the many assessment approaches can lead to different conclusions, they also can support and complement each other. For example, techniques that do not include adaptation tend to overestimate damages. However, by comparing the results from these methods with methods that do include adaptation, one determines the potential magnitude of adaptation. One can also compare results across space. If every method suggests one specific region is more sensitive to climate, there is increased confidence that this result must be true. Similarly, if all the results suggest another region is not sensitive to climate, then this increases confidence in that result as well.

Some methods are more appropriate for certain tasks than others. For example, crop modeling, production function approaches and intertemporal analyses are appropriate tools for measuring the dynamic short-term effects of weather or abrupt climate change. These methods capture examples where farmers are not going to be able to adapt. Consequently, these methods are not good predictors of long-term climate change precisely because they do not include adaptation. The impacts of long-term climate change are better captured by agro-economic models and Ricardian methods.

This review reveals another important conclusion. The agriculture sector includes many activities. In order to capture a broad picture of what will happen to the sector, it is important to model the entire sector. Thus it can be very misleading to study a single crop. Even expanding to a set of similar crops can lead to a biased picture if they are not representative of all crops. Finally, to get the complete picture, one must include not only crops but also livestock. Livestock has its own response to climate, and the livestock response does not necessarily mirror the crop response.

NOTES

1. This chapter builds on and extends previous reviews by Adams et al. (1998), Kurukulasuriya and Rosenthal (2003), Cline (2007) and Iglesias and Garrote (2007).
2. International Consortium for Application of Systems Approaches to Agriculture – International Benchmark Sites Network for Agro-technology Transfer.

REFERENCES

Adams, R.M., B.H. Hurd, S. Lenhart and N. Leary (1998), 'Effects of global climate change on agriculture: an interpretative review', *Climate Research*, **11**, 19–30.
Adams, R.M., R.A. Fleming, C.C. Chang, B.A. McCarl and C. Rosenzweig (1995), 'A reassessment of the economic effects of global climate change on US agriculture', *Climatic Change*, **30**(2), 147–67.
Adams, R.M., B.A. McCarl, K. Segerson, C. Rosenzweig, K.J. Bryant, B.L. Dixon, R. Conner, R.E. Evenson and D. Ojima (1999), 'Economic effects of climate change on US agriculture', in Robert Mendelsohn and James E. Neumann (eds), *The Impact of Climate Change on the United States Economy*, Cambridge, UK: Cambridge University Press.
Adams, R.M., C. Rosenzweig, R. Peart, J. Ritchie, B. McCarl, J. Glyer, B. Curry, J. Jones, K. Boote and L. Allen (1990), 'Global climate change and US agriculture', *Nature*, **345**, 219–24.
Aggarawal, P.K. and S.K. Sinha (1993), 'Effect of probable increase in carbon dioxide and temperature on wheat yields in India', *Journal of Agricultural Meteorology*, **48**(5), 811–14.

Alexandrov, V.A. and G. Hoogenboom (2000), 'The impact of climate variability and change on crop yield in Bulgaria', *Agricultural and Forest Meteorology*, **104**, 315–27.

Avila, F., L.J. Irias and M.A. de Lima (2006), 'Global warming effects on Brazilian agriculture: economic impact assessment on land values', Brazil: EMBRAPA.

Balti, Mohammed Nabil (2001), 'Evaluación económica de los effectos del cambio climático sobre la agricultura de Túnez y España', MSc thesis, Mediterranean Agronomic Institute of Zaragoza, June.

Barrow, E.M. and M.A. Semenov (1995), 'Climate change scenarios with high spatial and temporal resolution for agricultural application', *Forestry*, **68**, 349–60.

Bazzaz, F.A. and E.D. Fajer (1992), 'Plant life in a CO_2 rich world', *Scientific American*, **266**, 68–74.

Benhin, James (2007), 'Climate change and South African agriculture: impacts and adaptation options', CEEPA Discussion Paper No. 21, University of Pretoria, South Africa.

Chang, Ching-Cheng (2002), 'The potential impact of climate change on Taiwan's agriculture', *Agricultural Economics*, **27**(1), 51–64.

Chavas, J.P., K. Kim, J.G. Lauer, R.M. Klemme and W.L. Bland (2001), 'An economic analysis of corn yield, corn profitability, and risk at the edge of the Corn Belt', *Journal of Agricultural and Resource Economics*, **26**, 230–47.

Cline, William R. (1992), *The Economics of Global Warming*, Washington, DC, USA: Institute for International Economics.

Cline, William R. (1996), 'The impact of global warming on agriculture: comment', *American Economic Review*, **86**(5), 1309–11.

Cline, William R. (2007), 'Global warming and agriculture-impact estimates by country', Washington, DC, USA: Center for Global Development and Peterson Institute for International Economics.

Coll, Moshe and Lesley Hughes (2007), 'Effects of elevated CO_2 on an insect omnivore: a test for nutritional effects mediated by host plants and prey', *Agriculture Ecosystems & Environment*, (available online 6 August 2007).

CROPWAT (1995), 'A computer program for irrigation planning and management', FAO Irrigation and Drainage Paper No. 46, CROPWAT Version 5.6 (1991), Rome, Italy: FAO.

CROPWAT (2004), CROPWAT Version 7.0 software for Windows in English, French and Spanish, Water Resources, Development and Management Service, FAO, Rome, Italy.

Crosson, P. (1993), 'Impacts of climate change on the agriculture and economy of the Missouri, Iowa, Nebraska and Kansas (MINK) Region', in H. Kaiser and T. Drennen (eds), *Agricultural Dimensions of Global Climate Change*, Boca Raton, FL, USA: St. Lucie Press.

Darwin, Roy (1999), 'The impact of global warming on agriculture: a Ricardian analysis: comment', *American Economic Review*, **89**(4), 1049–52.

Darwin, Roy, Marinos Tsigas, Jan Lewandrowski and Anton Raneses (1995), 'World agriculture and climate change: economic adaptations', United States Department of Agriculture (USDA), Agricultural Research Service, Agricultural Economic Report 703, Washington, DC, USA: USDA.

Darwin, Roy, Marinos Tsigas, Jan Lewandrowski and Anton Raneses (1996), 'Land use and cover in ecological economies', *Ecological Economics*, **17**, 157–81.

Deressa, Temesgen Tadesse (2007), 'Measuring the economic impact of climate change on Ethiopian agriculture: Ricardian approach', World Bank Policy Research Working Paper 4342, available at http://econ.worldbank.org/resource.php?type=5.

Deschenes, O. and M. Greenstone (2007), 'The economic impacts of climate change: evidence from agricultural output and random fluctuations in weather', *American Economic Review*, **97**, 354–85.

Dinar, A., R. Hassan, R. Mendelsohn and J. Benhin (2008), *Climate Change and Agriculture in Africa: Impact Assessment and Adaptation Strategies*, London: Earthscan.

Dinar, A., R. Mendelsohn, R.E. Evenson, J. Parikh, A. Sanghi, K. Kumar, J. McKinsey and S. Lonergan (eds) (1998), 'Measuring the impact of climate change on Indian agriculture', World Bank Technical Paper 402, Washington, DC, USA: World Bank.

Easterling, W., P. Crosson, N. Rosenberg, M. McKenney, L. Katz and K. Lemon (1993), 'Agricultural impacts of and response to climate change in the Missouri-Iowa-Nebraska-Kansas (MINK) region', *Climatic Change*, **24**, 23–61.

Eid, Helmy M., Samia M. El-Marsafawy and Samiha A. Ouda (2007), 'Assessing the economic impacts of climate change on agriculture in Egypt: a Ricardian approach', World Bank Policy Research Working Paper 4293, available at http://econ.worldbank.org/resource.php?type=5.

El-Shaer, H.M., C. Rosenzeig, A. Iglesias, M.H. Eid and D. Hillel (1996), 'Impact of climate change on possible scenarios for Egyptian agriculture in the future', *Mitigation and Adaptation Strategies for Global Change*, **1**, 233–50.

FAO (1978), *Report on Agro-Ecological Zones; Volume 1: Methodology and Results for Africa*, Rome: FAO.

FAO (2004), FAOCLIM, a CD-ROM with world-wide agroclimatic data, available at www.fao.org/sd/2001/EN1102_en.htm.

Fleischer, A., I. Lichtman and R. Mendelsohn (2008), 'Climate change, irrigation, and Israeli agriculture: will warming be harmful?', *Ecological Economics*, **65**, 508–15.

Gbetibouo, G.A. and R.M. Hassan (2005), 'Measuring the economic impact of climate change on major South African field crops: a Ricardian approach', *Global and Planetary Change*, **47**(2–4), 143–52.

González, J.U. and Roberto Velasco (2008), 'Evaluation of the impact of climate change on the economic value of land in agricultural systems in Chile', *Chilean Journal of Agricultural Research*, **68**(1), 56–68.

Granados, J., I. Baquero and M.R. Gómez (2006), 'Effect of Global warming on Columbian agriculture', Columbia: INIA.

Guereña, Arantxa, Margarita Ruiz-Ramos, Carlos H. Díaz-Ambrona, José R. Conde and M. Inés Mínguez (2001), 'Assessment of climate change and agriculture in Spain using climate models', *Agronomy Journal*, **93**, 237–49.

Hansen, J.W., J.W. Jones, A. Irmak and F. Royce (2001), 'El Niño-Southern Oscillation impacts on crop production in the Southeast United States', in C. Rosenzweig, K.J. Boote, S. Hollinger, A. Iglesias and J. Phillips (eds), *Impacts of El Niño and Climate Variability on Agriculture*, ASA Special Publication No. 63, Madison, WI, USA: American Society of Agronomy, pp. 57–78.

Horie, T., J.T. Baker, H. Nakagawa and T. Matsui (2000), 'Crop ecosystem responses to climate change: rice', in K.R. Reddy and H.F. Hodges (eds), *Climate Change and Global Crop Productivity*, Wallingford, UK: CABI Press, pp. 81–106.

Iglesias, A. and M.I. Minguez (1996), 'Modelling crop–climate interactions in Spain. Vulnerability and adaptation of different agricultural systems to climate change', *Mitigation and Adaptation Strategies for Global Change*, **1**, 273–88.

Iglesias, Ana and Luis Garrote (2007), 'Projection of economic impacts of climate change in sectors of Europe based on bottom-up analysis (PESETA)', European Commission Joint Research Centre, Seville, Spain.

Iglesias, A., C. Rosenzweig and D. Pereira (1999), 'Prediction spatial impacts of climate in agriculture in Spain', *Global Environmental Change*, **10**, 69–80.

IPCC (Intergovernmental Panel on Climate Change) (2007a), *Climate Change 2007: The Physical Science Basis*, contribution of Working Group I to the Fourth Assessment Report of the Intergovernmental Panel on Climate Change, Cambridge, UK: Cambridge University Press.

IPCC (Intergovernmental Panel on Climate Change) (2007b), *Impacts, Adaptation and Vulnerability*, Cambridge, UK: Cambridge University Press.

Jain, Suman (2007), 'An empirical economic assessment of impacts of climate change on agriculture in Zambia', World Bank Policy Research Working Paper 4291, available at http://econ.worldbank.org/resource.php?type=5.

Jativa, Pablo and S. Niggol Seo (2006), 'A cross-sectional examination of climate change impacts on agriculture in Ecuador', report, Instituto Nacional Autónomo de Investigaciones Agropecuarias (INIAP), Ecuador.

Jin, Z., D. Ge, H. Chen and J. Fang (1994), 'Effects of climate change on rice production and strategies for adaptation in Southern China', in C. Rosenzweig and A. Iglesias (eds), *Implications of Climate Change for International Agriculture: Crop Modelling Study*, Washington, DC, USA: USEPA.

Johnston, T. and Q. Chiotti (2000), 'Climate change and the adaptability of agriculture: a review', *Journal of the Air & Waste Management Association*, **50**(4), 563–9.

Jones, J.W., J.W. Hansen, F.S. Royce and C.D. Messina (2000), 'Potential benefits of climate forecasting to agriculture', *Agricultural Ecosysem Environment*, **82**, 169–84.

Jorgenson, Dale W., Richard J. Goettle, Brian H. Hurd, Joel B. Smith, Lauraine G. Chestnut and David M. Mills (2004), *US Market Consequences of Global Climate Change*, Washington, DC, USA: Pew Center on Global Climate Change.

Kabubo-Mariara, Jane and Fredrick K. Karanja (2007), 'The economic impact of climate change on Kenyan crop agriculture: a Ricardian approach', World Bank Policy Research Working Paper 4334, available at http://econ.worldbank.org/resource.php?type=5.

Kaiser, H.M. and P. Crosson (1995), 'Implications of climate change for US agriculture', *American Journal of Agricultural Economics*, **77**(3), 734–40.

Kaiser, H.M., S.J. Riha, D.S. Wilks, D.G. Rossiter and R. Sampath (1993), 'A farm-level analysis of economic and agronomic impacts of gradual warming', *American Journal of Agricultural Economics*, **75**(2), 387–98.

Kapetanaki G. and C. Rosenzweig (1997), 'Impact of climate change on maize yield in Central and Northern Greece: a simulation study with CERES-maize', *Mitigation and Adaptation Strategies for Global Change*, **1**, 251–71.

Kelly, D.L., C.D. Kolstad and G.T. Mitchell (2005), 'Adjustment costs from environmental change', *Journal of Environmental Economics and Management*, **50**, 468–95.

Kimball, B.A., J.R. Mauney, F.S. Nakayama and S.B. Idso (1993), 'Effects of elevated CO_2 and climate variables on plants', *Journal of Soil and Water Conservation*, **48**, 9–14.

Krishnan, P., D.K. Swain, H. Chandra Bhaskar, S.K. Nayak and R.N. Dash (2007), 'Impact of elevated CO_2 and temperature on rice yield and methods of adaptation by crop simulation studies', *Agriculture Ecosystems & Environment*, **122**, 233–42.

Kumar, K.S.K. and Jyoti Parikh (1998), 'Climate change impacts on Indian agriculture: the Ricardian approach', in A. Dinar, R. Mendelsohn, R.E. Evenson, J. Parikh, A. Sanghi, K. Kumar, J. McKinsey and S. Lonergan (eds), *Measuring the Impact of Climate Change on Indian Agriculture*, World Bank Technical Paper 402. Washington, DC, USA: World Bank.

Kumar, K.S.K. and Jyoti Parikh (2001), 'Indian agriculture and climate sensitivity', *Global Environmental Change: Human and Policy Dimensions*, **11**(2), 147–54.

Kurukulasuriya, P. and M.I. Ajwad (2007), 'Application of the Ricardian technique to estimate the impact of climate change on smallholder farming in Sri Lanka', *Climatic Change*, **81**(1), 39–59.

Kurukulasuriya, P. and R. Mendelsohn (2008a), 'Crop switching as an adaptation strategy to climate change', *African Journal of Agriculture and Resource Economics*, **2**, 105–26.

Kurukulasuriya, P. and R. Mendelsohn (2008b), 'A Ricardian analysis of the impact of climate change on African cropland', *African Journal of Agriculture and Resource Economics*, **2**, 1–23.

Kurukulasuriya, P. and R. Mendelsohn (2008c), 'Modeling endogenous irrigation: the impact of climate change on farmers in Africa', World Bank Policy Research Working Paper 4278, Washington, DC, USA: World Bank.

Kurukulasuriya, P. and S. Rosenthal (2003), 'Climate change and agriculture', World Bank Environment Department Paper 91, Washington, DC, USA: World Bank.

Kurukulasuriya P., R. Mendelsohn, R. Hassan, J. Benhin, T. Deressa, M. Diop, H. Mohamed Eid, K. Yerfi Fosu, G. Gbetibouo, S. Jain, A. Mahamadou, R. Mano, J. Kabubo-Mariara, S. El-Marsafawy, E. Molua, S. Ouda, M. Ouedraogo, I. Sène, D. Maddison, S. Niggol Seo and A. Dinar (2006), 'Will African agriculture survive climate change?', *World Bank Economic Review*, **20**(3), 367–88.

Lal, R., H.M. Hassan and J. Dumanski (1999), 'Desertification control to sequester C and mitigate the greenhouse effect', in R.J. Rosenberg, R.C. Izaurrade and E.L. Malone (eds), *Carbon Sequestration in Soils: Science Monitoring and Beyond*, Columbus, OH, USA: Batelle Press, pp. 83–107.

Lanfranco, B. and J. Lozanoff (2006), 'Climate and rural poverty in Uruguay', report, Instituto Nacionale de Investigación Agropecuaria (INIA), Uruguay.

Lang, Gunter (2001), 'Global warming and German agriculture – impact estimations using a restricted profit function', *Environmental & Resource Economics*, **19**(2), 97–112.

Liu, Hui, Xiubin Li, Guenther Fischer and Laixiang Sun (2004), 'Study on the impacts of climate change on China's agriculture', *Climatic Change*, **65**(1–2), 125–48.

Long, S.P., E.A. Ainsworth, A.D.B. Leakey, J. Nösberger and D.R. Ort (2006), 'Food for thought: lower-than-expected crop yield stimulation with rising CO_2 concentrations', *Science*, **312**(5782), 1918–21.

Lozanoff, J. and E. Cap (2006), 'Impact of climate change over Argentine agriculture: an economic study', report, El Instituto Nacionale de Techología Agropecuaria (INIA), Argentina.

Maddison, D. (2000), 'A hedonic analysis of agricultural land prices in England and Wales', *European Review of Agricultural Economics*, **27**(4), 519–32.

Maddison, David, Marita Manley and Pradeep Kurukulasuriya (2007), 'The impact of climate change on African agriculture: a Ricardian approach', World Bank Policy Research Working Paper 4306, available at http://econ.worldbank.org/resource.php?type=5.

Mano, Reneth and Charles Nhemachena (2007), 'Assessment of the economic impacts of climate change on agriculture in Zimbabwe: a Ricardian approach', World Bank Policy Research Working Paper 4292, available at http://econ.worldbank.org/resource.php?type=5.

Mathews, J.T. (1991), *Greenhouse Warming: Negotiating a Global Regime*, Washington, DC, USA: World Resources Institute.

Mavromatis, T. and P.D. Jones (1998), 'Comparison of climate change scenario construction methodologies for impact assessment studies', *Agricultural Forest and Meteorology*, **91**, 51–67.

McKinsey, James W. Jr. and Robert E. Evenson (1998), 'Technology–climate interaction: was the green revolution in India climate-friendly?', in A. Dinar, R. Mendelsohn, R.E. Evenson, J. Parikh, A. Sanghi, K. Kumar, J. McKinsey and S. Lonergan (eds), *Measuring the Impact of Climate Change on Indian Agriculture*, World Bank Technical Paper 402, Washington, DC, USA: World Bank.

Mearns, L.O., C. Rosenzweig and R. Goldberg (1996), 'The effect of changes in daily and interannual climatic variability on CERES-wheat: a sensitivity study', *Climatic Change*, **32**(3), 257–92.

Mearns, L.O., T. Mavromatis, E. Tsvetsinskaya, C. Hays and W. Easterling (1999), 'Comparative responses of EPIC and CERES crop model to high and low spatial resolution climate change scenarios', *Journal of Geophysical Research*, **40**(D6), 6623–46.

Mendelsohn, R. and A. Dinar (1999), 'Climate change, agriculture, and developing countries: does adaptation matter?', *World Bank Research Observer*, **14**(2), 277–93.

Mendelsohn, R. and A. Dinar (2003), 'Climate, water, and agriculture', *Land Economics*, **79**(3), 328–41.

Mendelsohn, R. and W. Nordhaus (1996), 'The impact of global warming on agriculture: reply to Cline', *American Economic Review*, **86**(5), 1312–15.

Mendelsohn, R. and W. Nordhaus (1999a), 'The impact of global warming on agriculture: a Ricardian analysis: reply to Quiggin and Horowitz', *American Economic Review*, **89**(4), 1046–48.

Mendelsohn, R. and W. Nordhaus (1999b), 'The impact of global warming on agriculture: a Ricardian analysis: reply to Darwin', *American Economic Review*, **89**(4), 1053–55.

Mendelsohn, Robert and Reinsborough, Michelle (2007), 'A Ricardian analysis of US and Canadian farmland', *Climatic Change*, **81**(1), 9–17.

Mendelsohn, R. and L. Williams (2004), 'Comparing forecasts of the global impacts of climate change', *Mitigation and Adaptation Strategies for Global Change*, **9**, 315–33.

Mendelsohn, R., A. Dinar and A. Sanghi (2001), 'The effect of development on the climate sensitivity of agriculture', *Environment and Development Economics*, **6**(1), 85–101.

Mendelsohn, R., W.D. Nordhaus and D. Shaw (1994), 'The impact of global warming on agriculture: a Ricardian analysis', *American Economic Review*, **84**(4), 753–71.

Mendelsohn, R., W.D. Nordhaus and D. Shaw (1999), 'The impact of climate variation on US agriculture', in R. Mendelsohn and J. Neumann (eds), *The Impact of Climate Change on the United States Economy*, Cambridge, UK: Cambridge University Press.

Mendelsohn, R., William D. Nordhaus and Daigee Shaw (1996), 'Climate impacts on aggregate farm value: accounting for adaptation', *Agricultural and Forest Meteorology*, **80**, 55–66.

Mendelsohn, Robert, Antonio Flavio Dias Ávila and S. Niggol Seo (2007a), 'Incorporation of the climate change to the strategies of rural development: synthesis of the Latin America results', Montevideo, Uruguay: PROCISUR.

Mendelsohn, Robert, Pradeep Kurukulasuriya, Alan Basist, Felix Kogan and Claude Williams (2007b), 'Climate analysis with satellite versus weather station data', *Climatic Change*, **81**(1), 71–84.

Mendelsohn, R., A. Basist, A. Dinar and P. Kurukulasuriya (2007c), 'What explains agricultural performance: climate normals or climate variance?', *Climatic Change*, **81**, 85–99.

Mendelsohn, R., A. Basist, P. Kurukulasuriya and A. Dinar (2007d), 'Climate and rural income', *Climatic Change*, **81**(1), 101–118.

Mizina, S.V., J.B. Smith and E.F. Gossen (1997), 'Development of wheat management strategy taking into account possible climate change in Kazakhstan', *Hydrometeorology and Ecology*, **3**, 64–72.

Mizina, S.V., J.B. Smith, E. Gossen, K.F. Spiecker and S.L. Witkowski (1999), 'An evaluation of adaptation options for climate change impacts on agriculture in Kazakhstan', *Mitigation and Adaptation Strategies for Global Change*, **4**, 25–41.

Molua, Ernst, L. (2002), 'Climate variability, vulnerability and effectiveness of farm-level adaptation options: the challenges and implications for food security in Southwestern Cameroon', *Environment and Development Economics*, **7**(3), 529–45.

Molua, Ernest L. and Cornelius M. Lambi (2007), 'The economic impact of climate change on agriculture in Cameroon', World Bank Policy Research Working Paper 4364, available at http://econ.worldbank.org/resource.php?type=5.

Mooney, H.A. and W. Koch (1994), 'The impact of rising CO_2 concentrations on the terrestrial biosphere', *Ambio*, **23**(1), 74–6.

Mount, T. and Z. Li (1994), 'Estimating the effects of climate change on grain yield and production in the US', Washington, DC, USA: USDA, Economic Research Services.

Muchena, P. (1994), 'Implications of climate change for maize yields in Zimbabwe', in C. Rosenzweig and A. Iglesias (eds), *Implications of Climate Change for International Agriculture: Crop Modeling Study*, Washington, USA: US-EPA.

Muchena, P. and A. Iglesias (1995), 'Vulnerability of maize yields to climate change in different farming sectors in Zimbabwe', *American Society of Agronomy*, **59**, 229–39.

Nordhaus, W.D. (1991), 'To slow or not to slow: the economics of the greenhouse effect', *Economic Journal*, **101**, 920–37.

Onyeji, S.C. and G. Fischer (1994), 'An economic analysis of potential impacts of climate change in Egypt', *Global Environmental Change and Policy Dimensions*, **4**(4), 281–99.

Ouedraogo, Mathieu (1999), 'Contribution à l'évaluation de l'impact économique des changements climatiques sur l'agriculture Marocaine', MSc thesis, Institut Agronomique et Vétérinaire Hassan II, Rabat, Morocco (September).

Ouedraogo, Mathieu, Leopold Some and Youssouf Dembele (2007), 'Economic impact assessment of climate change on agriculture in Burkina Faso: a Ricardian approach', CEEPA Discussion Paper No. 24, University of Pretoria, South Africa.

Pacheco, R. (2006), 'Clima y Pobreza Rural: Incorporación del Clima a las Estrategias de Desarollo Rural', report, Instituto Nacionale de Investigaciones Agrícolas (INIA), Venezuela.

Parry, M.L. and C. Rosenzweig (1993), 'Food supply and risk of hunger', *Lancet*, **342**, 1345–7.

Parry, M.L., C. Rosenzweig, A. Iglesias, M. Livermore and G. Fischer (2004), 'Effects of climate change on global food production under SRES emissions and socio-economic scenarios', *Global Environmental Change*, **14**, 53–67.

Parry, Martin, Cynthia Rosenzweig and Matthew Livermore (2005), 'Climate change, global food supply and risk of hunger', *Philosophical Transactions of the Royal Society B*, **360**, 2125–38, published online, 24 October.

Pearce, D.W., W.R. Cline, A.N. Achanta, S. Fankhauser, R.K. Pachauri, R.S.J. Toll and P. Vellinga (1996), 'The social costs of climate change: Greenhouse damage and the benefits of control', in J.P. Bruce, H. Lee and E.F. Haites (eds), *Climate Change 1995: Economic and Social Dimensions*, contribution of Working Group III to the Second Assessment Report of the Intergovernmental Panel on Climate Change, Cambridge, UK: Cambridge University Press, pp. 179–224.

Plantinga, A.J. and D.J. Miller (2001), 'Agricultural land values and the value of rights to future land development', *Land Economics*, **77**(1), 56–67.

Polsky, C. (2004), 'Putting space and time in Ricardian climate change impact studies: agriculture in the US Great Plains, 1969–1992', *Annals of the Association of American Geographers*, **94**(3), 549–64.

Polsky, C. and W.E. Easterling (2001), 'Adaptation to climate variability and change in the US Great Plains: a multi-scale analysis of Ricardian climate sensitivities', *Agriculture Ecosystems & Environment*, **85**, 133–44.

Quiggin, John and John K. Horowitz (1999), 'The impact of global warming on agriculture: a Ricardian analysis: comment', *American Economic Review*, **89**(4), 1044–5.

Reilly, J., F.N. Tubiello, B. McCarl and J. Melillo (2001), 'Impacts of climate change and variability on agriculture', in US National Assessment of the Potential Consequences of Climate Variability and Change, Washington, DC: US Global Change Research Program.

Reinsborough, M.J. (2003), 'A Ricardian model of climate change in Canada', *Canadian Journal of Economics – Revue Canadienne D'Economique*, **36**(1), 21–40.

Ritchie, J.T., U. Singh, D.C. Godwin and W.T. Bowen (1998), 'Cereal growth, development and yield', in G.Y. Tsuji, G. Hoogenboom and P.K. Thornton (eds), *Understanding Options for Agricultural Production*, Dordrecht, The Netherlands: Kluwer Academic Publishers, pp. 79–98.

Rosenzweig, C. and A. Iglesias (1998), 'The use of crop models for international climate change impact assessment', in G.Y. Tsuji, G. Hoogenboom and P.K. Thornton (eds), *Understanding Options for Agricultural Production*, Dordrecht, The Netherlands: Kluwer Academic Publishers, pp. 267–92.

Rosenzweig, C. and F.N. Tubiello (1996), 'Impacts of global climate change on Mediterranean agriculture: current methodologies and future directions', *Mitigation and Adaptation Strategies for Global Change*, **1**, 219–32.

Rosenzweig, C. and M.L. Parry (1994), 'Potential impact of climate change on world food supply', *Nature*, **367**, 133–8.

Rosenzweig, C., A. Iglesias and M.L. Parry (1993), 'Potential impacts of climate change on world food supply: a summary of a recent international study', in H.M. Kaiser and T.E. Drennen (eds), *Agricultural Dimensions of Global Climate Change*, Delray Beach, FL, USA: St. Lucie Press, pp. 87–116.

Rosenzweig, C., A. Iglesias, G. Fischer, Y. Liu, W. Baethgen and J.W. Jones (1999), 'Wheat yield functions for analysis of landuse change in China', *Environmental Modeling and Assessment*, **4**, 128–32.

Sands, R.D. and J.A. Edmonds (2005), 'Climate change impacts for the conterminous USA: an integrated assessment – Part 7. Economic analysis of field crops and land use with climate change', *Climatic Change*, **69**, 127–50.

Sanghi, A. and R. Mendelsohn (2008), 'The impacts of global warming on farmers in Brazil and India', *Global Environmental Change*, **18**(4), 655–65.

Sanghi, A., R. Mendelsohn and A. Dinar (1998), 'The climate sensitivity of Indian agriculture', in A. Dinar, R. Mendelsohn, R.E. Evenson, J. Parikh, A. Sanghi, K. Kumar, J. McKinsey and S. Lonergan (eds), *Measuring the Impact of Climate Change on Indian Agriculture*, World Bank Technical Paper 402, Washington, DC, USA: The World Bank.

Sanghi, A., D. Alves, R. Evenson and R. Mendelsohn (1997), 'Global warming impacts on Brazilian agriculture: estimates of the Ricardian model', *Economia Aplicada*, **1**(1), 7–33.

Saseendran, S.A., K.K. Singh, L.S. Rathore, S.V. Singh and S.K. Sinha (2000), 'Effects of climate change on rice production in the tropical humid climate of Kerala, India', *Climatic Change*, **44**, 495–514.

Scheraga, J.D., N.A. Leary, R.J. Goettle, D.W. Jorgenson and P.J. Wilcoxen (1993), 'Macroeconomic modeling and the assessment of climate change impacts', in Y. Kaya, N. Nakicenovic, W. Nordhaus and F.L. Toth (eds), *Costs, Impacts and Possible Benefits of CO$_2$ Mitigation*, International Institute for Applied Systems Analysis, Collaborative Paper Series, Laxenburg, Austria: IIASA, CP-93-2, pp. 107–32.

Schimel, D. (2006), 'Climate change and crop yields: beyond Cassandra', *Science*, **312**(5782), 1889–90, 30 June.

Schlenker, W., W.M. Hanemann and A.C. Fisher (2005), 'Will US agriculture really benefit from global warming? Accounting for irrigation in the hedonic approach', *American Economic Review*, **95**(1), 395–406.

Schlenker, W., W.M. Hanemann and A.C. Fisher (2006), 'The impact of global warming on US agriculture: an econometric analysis of optimal growing conditions', *Review of Economics and Statistics*, **81**(1), 113–25.

Schlenker, W., W.M. Hanemann and A.C. Fisher (2007), 'Water availability, degree days, and the potential impact of climate change on irrigated agriculture in California', *Climatic Change*, **81**(1), 19–38.

Sene, Isidor Marcel, Mbaye Diop and Alioune Dieng (2007), 'Impacts of climate

change on the revenues and adaptation of farmers in Senegal', CEEPA
Discussion Paper No. 20, University of Pretoria, South Africa.

Seo, S.N., R. Mendelsohn and M. Munasinghe (2005), 'Climate change and
agriculture in Sri Lanka: a Ricardian valuation', *Environment and Development
Economics*, **10**(5), 581–96.

Seo, S.N. and R. Mendelsohn (2008a), 'Climate change impacts and adaptations
on animal husbandry in Africa', *African Journal of Agriculture and Resource
Economics*, **2**, 65–82.

Seo, S.N. and R. Mendelsohn (2008b), 'A Ricardian analysis of the impact of
climate change on South American farms', *Chilean Journal of Agricultural
Research*, **68**(1), 69–79.

Seo, S.N. and R. Mendelsohn (2008c), 'Measuring impacts and adaptation to
climate change: a structural Ricardian model of African livestock management',
Agricultural Economics, **38**, 150–65.

Seo, S.N. and R. Mendelsohn (2008d), 'An analysis of crop choice: adapting to
climate change in Latin American farms', *Ecological Economics*, **67**(1), 109–16.

Seo, S.N., R. Mendelsohn, A. Dinar, R. Hassan and P. Kurukulasuriya (2008a),
'A Ricardian analysis of the distribution of climate change impacts on agri-
culture across agro-ecological zones in Africa', *Environmental and Resource
Economics* (forthcoming).

Seo, S.N., R. Mendelsohn, P. Kurukulasuriya and A. Dinar (2008b), 'An analysis
of adaptation to climate change in African livestock management by agro-
ecological zones', *B.E. Journal of Economic Analysis & Policy* (forthcoming).

Sharpley, A.N. and J.R. Williams (eds) (1990), 'EPIC – erosion/productivity
impact calculator: 1. Model documentation', US Department of Agriculture
Technical Bulletin No. 1768.

Smith, J.B. and D. Tirpak (eds) (1989), *The Potential Impacts of Global Climate
Change on the United States*, Washington, DC: US Environmental Protection
Agency, EPA-230-05-X9-050.

Strzepek, Kenneth and Alyssa McCluskey (2007), 'The impacts of climate change
on regional water resources and agriculture in Africa', World Bank Policy
Research Working Paper 4290, Washington, DC, USA: World Bank.

Stuczyinski, T., G. Demidowicz, T. Deputat, T. Górski, S. Krazowicz and J.
Kus (2000), 'Adaptation scenarios of agriculture in Poland to future climate
changes', *Environmental Monitoring and Assessment*, **61**, 133–44.

Tao, F., M. Yokozawa, Y. Hayashi and E. Lin (2003), 'Future climate change,
the agricultural water cycle, and agricultural production in China', *Agriculture,
Ecosystems & Environment*, **95**, 203–15.

Tao, F., M. Yokozawa, Y. Xu, Y. Hayashi and Z. Zhang (2006), 'Climate
change and trends in phenology and yields of field crops in China, 1981–2000',
Agricultural and Forest Meteorology, **138**, 82–92.

Tol, R. (2002), 'Estimates of the damage costs of climate change', *Environmental
and Resource Economics*, **21**, 47–73.

Timmins, C. (2006), 'Endogenous land use and the Ricardian valuation of climate
change', *Environmental and Resource Economics*, **33**(1), 119–42.

Tubiello, Francesco N. and Frank Ewert (2002), 'Simulating the effect of elevated
CO_2 on crops: approaches and applications for climate change', *European
Journal of Agronomy*, **18**, 57–74.

Wahaj, Robina, Florent Maraux and Giovanni Munoz (2007), 'Actual crop water
use in project countries: a synthesis at the regional level', World Bank Policy

Research Working Paper 4288, available at http://econ.worldbank.org/resource.
php?type=5.

Wang, J., R. Mendelsohn, A. Dinar, J. Huang and S. Rozzelle (2008), 'Can China
continue feeding itself? The impact of climate change on agriculture', World
Bank Policy Research Working Paper 4470, available at http://econ.worldbank.
org/resource.php?type=5.

Weber, M. and G. Hauer (2003), 'A regional analysis of climate change impacts
on Canadian agriculture', *Canadian Public Policy – Analyse de Politiques*, **29**(2),
163–80.

Winters, P., R. Murgai, E. Sadoulet, A. De Janvry and G. Frisvold (1998),
'Economic and welfare impacts of climate change on developing countries',
Environmental and Resource Economics, **12**(1), 1–24.

Woodward, F.I. (1993), 'Leaf responses to the environment and extrapolation to
larger scales', in A. Solomon and H. Shugart (eds), *Vegetation Dynamics and
Global Change*, New York, USA: Chapman and Hall, pp. 71–100.

Wu, D., Q. Yu, C. Lu and H. Hengsdijk (2006), 'Quantifying production poten-
tials of winter wheat in the North China plain', *European Journal of Agronomy*,
24, 226–35.

Xiong, W., R. Matthews, I. Holman, E. Lin and Y. Xu (2007), 'Modelling China's
potential maize production at regional scale under climate change', *Climatic
Change*, **85**(3–4), 433–51.

Yang, X., E. Lin, S. Ma, H. Ju, L. Guo, W. Xiong, Y. Li and Y. Xu (2007),
'Adaptation of agriculture to warming in Northeast China', *Climatic Change*,
84, 45–58.

Yao, F., Y. Xu, E. Lin, M. Yokozawa and J. Zhang (2007), 'Assessing the impacts
of climate change on rice yields in the main rice areas of China', *Climatic
Change*, **80**, 395–409.

You, S.C. (2001), 'Agricultural adaptation to climate change in China', *Journal of
Environmental Sciences*, **13**(2), 192–7.

4. The Ricardian method

The traditional Ricardian method (Mendelsohn et al., 1994) examines how the value of farm land varies across a set of exogenous variables (such as climate and soils). The model assumes that farmers, given these exogenous constraints that they cannot control, choose a set of outputs and inputs to maximize profits. By regressing land value on these exogenous variables, using a cross-sectional analysis, the Ricardian method measures how these variables affect land value. Because farmers are adjusting inputs and outputs to match local conditions, the Ricardian method implicitly captures adaptation. However, because the adjustments are not explicitly modeled, the technique treats adaptation as a 'black box'. It does not reveal the explicit adjustments that are being made by individual farmers to take advantage of the conditions they face. This chapter reviews the foundation of the traditional Ricardian model in detail and discusses critical issues raised about the model. The chapter then discusses modifications that have been made to the model to make it suitable for studying developing countries.

TRADITIONAL RICARDIAN MODEL

The Ricardian method was named after Ricardo because of his original observation that in a competitive market, land rents would reflect the net revenue of farmland (Ricardo, 1817). Farmland value (V) consequently reflects the present value of the future stream of income from a parcel of land (measured as net revenue per hectare).[1] This principle is captured in the following equation:

$$V = \int P_{LE} e^{-\varphi t} \mathrm{d}t = \int \left[\sum P_i Q_i(X, F, Z) - \sum RX \right] e^{-\varphi t} \mathrm{d}t \quad (4.1)$$

where P_{LE} is the net revenue per hectare, P_i is the market price of crop i, Q_i is output of crop i, F is a vector of climate variables, Z is a vector of soil and economic variables, X is a vector of purchased inputs (other than land), R is a vector of input prices, t is time, and φ is the discount rate (Mendelsohn et al., 1994).

We assume that the farmer chooses Q and X to maximize net revenues given the characteristics of the farm and market prices. That is, Q and X are determined so as to maximize V in equation (4.1) for any given combination of exogenous variables. The resulting profit maximizing outcome is a reduced form model that examines how V is affected by the set of exogenous variables F, Z and P:

$$V = f(F, Z, P) \tag{4.2}$$

The standard Ricardian model is a quadratic formulation of climate and a linear function of all the other variables:

$$V = B_0 + B_1F + B_2F^2 + B_3Z + B_5P + u \tag{4.3}$$

where u is an error term. The linear and quadratic term for temperature and precipitation are introduced to capture known non-linearities in the response of crops to climate. For example, experiments with crops in laboratory settings suggest they tend to have hill-shaped response functions with respect to temperature (Chapter 2).

Differentiating equation (4.3) with respect to a climate component, f_i, (such as summer temperature or spring precipitation), yields:

$$dV/df_i = b_{1i} + 2 \cdot b_{1i} \cdot f_i \tag{4.4}$$

The marginal value of each climate component depends on both the linear and quadratic climate coefficients from (4.3). The marginal value also depends upon the level of the climate component, f_i. However, this functional form assumes that the marginal effect of climate is independent of all the other variables. The marginal value can be computed at the mean of the sample by substituting $E[f_i]$ into equation (4.4) for f_i. Alternatively, one can calculate the marginal value for each observation by using equation (4.4) and then calculating the mean of this distribution. With the linear model, the annual marginal value of temperature or precipitation is the sum of the marginal values across all four seasons.

An alternative functional form for the Ricardian model is loglinear:

$$\text{Ln } V = B_0 + B_1F + B_2F^2 + B_3Z + B_4G + B_5P + u \tag{4.5}$$

In this case, the effect of climate is not additive but multiplicative. Taking the derivative of equation (4.5) with respect to a climate variable f_i yields:

$$dV/df_i = V \cdot (b_{1i} + 2 \cdot b_{1i} \cdot f_i) \tag{4.6}$$

The sign of the marginal climate impact still depends upon the signs of the climate coefficients. However, the size of the marginal climate impact depends not only on climate but also on every other variable that determines land value.

The estimated Ricardian model can be used to value climate change. The welfare change, W, from a change in climate from some initial level $F0$ to a new level $F1$ is:

$$W = \sum [V(F1_i, Z_i, P_i) - V_i(F0_i, Z_i, P_i)] \cdot L_i \qquad (4.7)$$

where L_i is the amount of land in farm i. Each farm may have a different reaction to climate change. If the value of the farm increases in the new climate, the climate change is beneficial, whereas if it decreases, the climate change is harmful. By summing these effects across all observations and weighting by the amount of land in each farm, one can see the aggregate net effect. By examining these effects across a landscape, one can also detect how climate changes affect different locations.

In order to apply the model, the researcher must decide how to characterize climate. The Ricardian analyses done to date have largely focused on seasonal climate normals. A climate normal is the average of weather over 30 years and is a standard measure used by climatologists (National Climate Data Center). In principle, the mean values of temperature and precipitation for every month could be included so that one can capture the full effect of precipitation and temperature across the year. In practice, however, adjacent months are highly correlated with each other and one gets nonsensical monthly results. A successful alternative is to include seasonal measures, the average of three months. The vectors of temperature and precipitation include linear and quadratic terms for winter, spring, summer and autumn. It is also possible to include other measures of climate such as wind speed or sunlight but one must be careful about variables such as evapotranspiration that already reflect precipitation and temperature to avoid double counting effects.

An alternative measure of climate used in many agronomic models concerns only the growing season. However, there are several problems with using only the growing season in the Ricardian model. First, the model captures the effect of climate across several crops, and different crops, do not have consistent growing seasons. Second, growing seasons are endogenous. When farmers plant and harvest it is a choice that they can influence. It is likely that climate change will alter that choice. Third, the climate in periods when crops are not growing can affect farm net revenues. Cold winters, for example, can reduce pests and weeds (Chapter 2), lowering the cost of farming. Precipitation before the growing season can increase soil moisture, making it possible to plant earlier.

In addition to climate normals, some studies have included climate variance (Mendelsohn et al., 1999; Mendelsohn et al., 2007b). These studies have found that diurnal variance, the change in temperature over the day, and interannual variance, the change in weather from year to year, are both significant when included in a Ricardian model. Further, including the variance terms affects the climate normal coefficients. Finally, another alternative is to use degree days instead of temperature (Schlenker et al., 2006). This analysis carefully examines the shape of the relationship between degree days and land value in the US. However, this analysis has made no distinction between the effects of degree days in each season. Given both the laboratory and Ricardian results suggesting that climate has a different impact on plants and especially crops in their different stages of development, this is probably a misspecification.

The initial application of the Ricardian model examined farmland values across counties in the United States (Mendelsohn et al., 1994). The model accounts for precipitation and temperature. However, several authors have criticized this first paper for not explicitly modeling irrigation (Cline, 1996; Darwin, 1999; Schlenker et al., 2005). Agronomic research suggests that irrigated crops have a different response to climate than rainfed crops.

Schlenker et al. (2005) consequently proposed that rainfed crops and irrigated crops be analyzed separately. These authors show that the Ricardian response function of rainfed crops is different from the Ricardian response function of all crops in the US. The analysis revealed that rainfed crops were more sensitive to both temperature and rainfall than all crops together. The authors argued that this was proof that the original Ricardian model was biased. However, the original Ricardian model was the locus of land values across all crops. Even if rainfed and irrigated crops have different Ricardian response functions, this is not proof that an analysis that includes them both is biased. The analysis of all crops still yields an unbiased prediction of the effect of climate on all crops. Similarly, the analysis of just rainfed crops yields an unbiased prediction of the effect of climate on rainfed crops. Appropriate techniques for modeling rainfed and irrigated land separately in the Ricardian context are dealt with in Chapter 6.

Although it is not clear that the Ricardian model must explicitly include irrigation, it is important that the model take account of the availability of water sources from outside the county. County precipitation and temperature will capture water available from local sources. However, some farmers are in counties near rivers. These counties can take advantage of rainfall from far away counties through surface water withdrawals. Mendelsohn and Dinar (2003) examine whether these external sources of water are important in the US, and Fleischer et al. (2008) examine water withdrawals in Israel. Both studies find that external water substantially increases farm value. Including external water sources in the Ricardian model can alter the coefficients on climate. Ricardian studies that do not capture the availability of external water sources can be biased.

The model described above assumes that input and output prices are included in the Ricardian model. However, in many applications, the prices are the same across the data set because the farmers are all in the same market. In this case, the Ricardian method cannot discern the independent effects of prices, and price cannot be included (Cline, 1996). The omission of prices leads the Ricardian model to over-predict damages and benefits (Mendelsohn and Nordhaus, 1996). If production falls, prices will rise and soften the blow to farmers. Similarly, if production increases, prices will fall and farmers will not obtain quite the same gain. Prices for many food items (and especially staple foods), however, are rarely determined just by local conditions. Trading tends to make food prices a global issue. How climate affects food prices depends on the magnitude of the gains in some regions versus the losses in others (Reilly et al., 1994). Until a reasonable global model is constructed that quantifies climate impacts across the world, it is not clear what will happen to food prices in different climate scenarios. In contrast, it may be possible to predict how local productivity changes might affect the price of local inputs such as labor. Dramatic regional reductions in productivity are likely to depress local wage rates and reduce the welfare damages from climate change.

Another variable that is exogenous to the farmer is the level of carbon dioxide (CO_2). Higher CO_2 levels increase crop productivity and make crops less vulnerable to drought (Reilly et al., 1996). However, because CO_2 is distributed evenly across the planet, all farms in a cross-sectional data set face the same level of CO_2. The Ricardian method consequently cannot measure the beneficial effects of carbon fertilization. The effects of carbon fertilization need to be added to the welfare calculations using exogenous sources of information such as laboratory studies (Reilly et al., 1996).

Another concern raised about the Ricardian models is whether they capture the cost of transitioning from one farming method to another (Kaiser et al., 1993a, 1993b; Quiggin and Horowitz, 1999; Kelly et al., 2005). Farmers may have trouble identifying the new climate, they may have to retire equipment prematurely, they may face capital constraints, or they may not have the knowledge to farm under new conditions. The Ricardian method is a comparative static analysis. The method measures the long-run adaptation costs of farming under new climate conditions, but it does not measure the transition (dynamic) costs. Mendelsohn and Nordhaus (1999) argue that these transition costs are likely to be small because of the slow progress of climate change and the short lifetime of farm capital. However, the magnitude of the transition costs is not something that the Ricardian model can measure directly.

MODIFICATIONS FOR DEVELOPING COUNTRY APPLICATIONS

The Ricardian method was originally applied to a developed country (US). In order to adapt the Ricardian method to developing countries, a number of modifications had to be made in order to address problems specific to agriculture in these countries. One problem in many developing countries is the absence of sufficient meteorological stations, especially in rural areas. Weather has to be interpolated between weather stations using a weighted regression for each farm or county. The regression effectively creates a predicted weather surface over space using the station observations. A specific measure of climate, f_i, is regressed over a quadratic combination of latitude (LAT), longitude ($LONG$) and altitude (ALT):

$$f_i = a_0 + a_1 LAT + a_2 LAT^2 + a_3 LONG + a_4 LONG^2$$

$$+ a_5 ALT + a_6 ALT^2 \tag{4.8}$$

The observations include the weather stations that are within a reasonable distance of the farm in question (for example within 500 km). Because nearby stations are likely to be more relevant than distant stations, the observations are weighted by proximity or 1/distance. The predicted value from the regression for the altitude, latitude and longitude of the farm is the interpolated value of that climate variable. The interpolation needs to be repeated for each climate variable and each farm. The weather at a particular location can be inferred from this surface.

An alternative source of climate data is satellites. Satellites have the advantage that they can scan over the entire landscape. However, satellite measures are limited to what they can measure. For example, satellites can measure surface temperature but they cannot directly measure precipitation. Satellites can measure soil wetness, however, which is highly correlated with precipitation. In order to assess whether interpolated weather station or satellite data were preferred, they were both tested in Brazil, India and the US (Mendelsohn et al., 2007b). Each climate measure was introduced in a Ricardian regression so that the effectiveness of each measure in predicting farm values could be compared. The satellite temperature data was able to provide a more significant signal for crop growth than the weather station data. However, the precipitation data from weather stations proved to be more effective than soil wetness. Soil wetness suffered from distortions from water bodies (especially the coast), reflections from forests, and irrigation. The results suggest that the best combination was to use satellite measures of temperature and interpolated weather station data of precipitation.

Another major stumbling block for applying the Ricardian method in developing countries is the absence of reliable measures of farm land value. Many developing countries do not have private property rights for land. Land is held in common in villages or belongs to the state. In these areas, there are no land sales and so there is no recorded measure of the value of land. In order to overcome this problem, the Ricardian method can be modified to use net revenue per hectare. Net revenue per hectare is gross revenue minus costs. Net revenue per hectare (P_{LE}) can be regressed upon the same exogenous variables as in equation (4.3):

$$P_{LE} = B_0 + B_1F + B_2F^2 + B_3Z + B_4G + u \qquad (4.9)$$

It is the annual equivalent to equation (4.3). As an annual measure, net revenue per hectare has the disadvantage of reflecting annual weather, not long-term climate. Further, one cannot explore the importance of climate variance with one year of net revenue data. However, one of the disadvantages of land value is that it reflects long-term uses of the land including development value. Consequently, farmland near major urban areas may have high land values even when it does not have high annual net revenue.

The reliance on net revenue, however, introduces yet another problem. There are few existing data sets that measure the net revenue per hectare. Farm data that does exist largely measures gross revenue (aggregate yields). In order to implement the analysis in countries without land value data, it is necessary to collect new data using surveys of farms. Unlike typical surveys that collect data in limited geographic regions, the Ricardian technique requires that the sample should come from across very broad geographic zones so that farms could be observed across different climates. The studies have to be based on either very large countries (for example, Brazil, India, or China) or across many small ones (for example several countries spread across Africa or South America). Studies based in very small countries, such as Sri Lanka, must rely on a very simple description of climate such as annual temperature or precipitation (Seo et al., 2005).

The fact that net revenue is calculated by subtracting annual costs from annual gross revenue can create some difficulties. For many farmers, they grow more than one crop per year. They may have different plots each devoted to a crop or they may have more than one cropping season. It may be difficult to allocate a fixed cost such as the cost of machinery to each parcel of land.

The gross crop revenue per hectare is the sum of the quantities of crops grown times their price divided by the cropland area. This may be an average value for a heterogeneous farm, not an accurate measure of each plot. The relevant price is the price received by the farmer. This can be difficult to measure because some farmers may sell directly to a buyer from their farm and others may take their products to market. If one uses prices at markets, it is important to subtract the cost of transporting, preparing and selling goods at the market.

Another problem that pertains to developing countries is that farmers do not always sell all of their output. In some cases, a large fraction of output is consumed by the household. In order to be consistent with commercial farms, own consumption is valued at market prices. Although this is not likely to bias the results, it may underestimate the value of crops in remote locations where farmers may get low prices for selling in distant markets but receive high value from local food consumption.

It is relatively easy to quantify the cost of variable inputs such as seed and fertilizer. However, it is also important to measure the annual cost of farm capital (animal power, machinery and farm buildings). Assuming that a unit of capital costs K to purchase and it lasts T years, the annual rent, R, for that capital is:

$$R = \frac{rK}{1 - e^{-\varphi T}} \tag{4.10}$$

By collecting purchase cost data and gathering estimates of the lifetime of each piece of equipment, it is possible to convert capital costs to an annual cost. Of course, one difficulty with capital costs is that it may be difficult to know how to allocate these costs across hectares. For example, a farmer may have several plots or grow several crops. Some capital may be specific to a plot or crop, whereas other capital may be used in all plots. It consequently can be easier to measure farm net revenue than the net revenue of specific plots.

Another important issue in developing country farms concerns valuing own labor. A large number of small farms rely heavily on household members to provide labor. There is no observed wage for household labor. If their time is valued at the wage paid for hired help, many of these small farms would appear to consistently lose money. Subtracting the household labor at market value would lead to losses. It is obvious that household labor is valued at less than the wage paid for hired help. Hired help is often used only at critical times such as at harvest when a great deal of work must be done quickly. The hired wage rates during harvests are higher than the wages during the rest of the year. It is consequently not obvious what wage rate to assign to household labor. The cost of household labor is consequently not included in net revenue. The size of the household was included to capture the supply of labor, but this is clearly imperfect. The omission of household labor costs may consequently suggest that small farms earn slightly higher net revenue per hectare than commercial farms. Omitting household labor costs from net revenues, however, is not expected to cause the climate change measures to be biased.

One important component of agriculture that should not be overlooked is livestock. Livestock in developed countries may be relatively climate-neutral because many animals are housed in barns and sheds and fed grain and feedstock collected elsewhere (Adams et al., 1999; Chapter 2). However, livestock in developing countries are kept outdoors and often graze on natural lands. There is every reason to expect that livestock are vulnerable to climate through direct exposure, through disease vectors, and through changes in ecosystems that affect grazing potential. In order to measure livestock net revenues it is important to measure outputs, inputs and capital. Outputs can be animals sold but animal products such as milk, meat, eggs and hides are also very important. Inputs tend to be young animals so it is important to place a different price on young (input) versus mature (output) animals. Capital tends to be the stock itself, the inventory of animals owned.

In many countries, animals are grazed on private land so it is relatively easy to measure the net revenue per hectare. However, in some locations, animals are raised on public or common lands. Even with a farm survey, it is very difficult to measure the amount of land each farmer uses for grazing in commons. With common land, it may not be possible to measure net revenue per hectare. The analysis may have to resort to measuring net revenue per farm. One alternative in this case is to develop a two-equation model that first predicts the number of animals owned per farm and then predicts the net revenue per animal owned.

NOTE

1. Net revenue is calculated as the (gross) revenue from sale of products minus costs. Depending on the cost categories included in this calculation, one can end up with several net revenue values.

REFERENCES

Adams, R.M., B.A. McCarl, K. Segerson, C. Rosenzweig, K.J. Bryant, B.L. Dixon, R. Conner, R.E. Evenson and D. Ojima (1999), 'Economic effects of climate change on US agriculture', in R. Mendelsohn and J. Neumann (eds), *The Impact of Climate Change on the United States Economy*, Cambridge, UK: Cambridge University Press.

Cline, William (1996), 'The impact of global warming on agriculture: comment', *American Economic Review*, **86**, 1309–12.

Darwin, Roy (1999), 'Comment on the impact of global warming on agriculture: a Ricardian analysis', *American Economic Review*, **89**, 1049–52.

Fleischer, A., I. Lichtman and R. Mendelsohn (2008), 'Climate change, irrigation, and Israeli agriculture: will warming be harmful?', *Ecological Economics*, **67**, 109–16.

Kaiser, H.M., S.J. Riha, D.S. Wilkes and R.K. Sampath (1993a), 'Adaptation to global climate change at the farm level', in H. Kaiser and T. Drennen (eds), *Agricultural Dimensions of Global Climate Change*, Delray Beach, FL, USA: St. Lucie Press.

Kaiser, H.M., S.J. Riha, D.S. Wilkes, D.G. Rossiter and R.K. Sampath (1993b), 'A farm-level analysis of economic and agronomic impacts of gradual warming', *American Journal of Agricultural Economics*, **75**, 387–98.

Kelly, D.L., C.D. Kolstad and G.T. Mitchell (2005), 'Adjustment costs from environmental change', *Journal of Environmental Economics and Management*, **50**, 468–95.

Mendelsohn, R. and A. Dinar (2003), 'Climate, water, and agriculture', *Land Economics*, **79**, 328–41.

Mendelsohn, R. and W. Nordhaus (1996), 'The impact of global warming on agriculture: reply to Cline', *American Economic Review*, **86**, 1312–15.

Mendelsohn, R. and W. Nordhaus (1999), 'The impact of global warming on agriculture: reply to Quiggin and Horowitz', *American Economic Review*, **89**, 1046–48.

Mendelsohn, R., W. Nordhaus and D. Shaw (1994), 'Measuring the impact of global warming on agriculture', *American Economic Review*, **84**, 753–71.

Mendelsohn, R., W. Nordhaus and D. Shaw (1999), 'The impact of climate variation on US agriculture', in R. Mendelsohn and J. Neumann (eds), *The Impact of Climate Change on the United States Economy*, Cambridge, UK: Cambridge University Press, pp. 55–74.

Mendelsohn, R., A. Basist, A. Dinar, F. Kogan, P. Kurukulasuriya and C. Williams (2007a), 'Climate analysis with satellites versus weather station data', *Climatic Change*, **81**, 71–84.

Mendelsohn, R., A. Basist, A. Dinar and P. Kurukulasuriya (2007b), 'What explains agricultural performance: climate normals or climate variance?', *Climatic Change*, **81**, 85–99.

Quiggin, J. and J.K. Horowitz (1999), 'The impact of global warming on agriculture: a Ricardian analysis: comment', *American Economic Review*, **89**, 1044–5.

Reilly, J., N. Hohmann and S. Kane (1994), 'Climate change and agricultural trade: who benefits, who loses?', *Global Environmental Change*, **4**, 24–36.

Reilly, J. et al. (1996), 'Agriculture in a changing climate: impacts and adaptations', in IPCC, *Climate Change 1995: Impacts, Adaptations, and Mitigation of Climate Change*, Cambridge, UK: Cambridge University Press, pp. 427–68.

Ricardo, David (1817), *On the Principles of Political Economy and Taxation*, London: John Murray.

Schlenker, W., M. Hanemann and A. Fischer (2005), 'Will US agriculture really benefit from global warming? Accounting for irrigation in the hedonic approach', *American Economic Review*, **95**(1), 395–406.

Schlenker, W., M. Hanemann and A. Fischer (2006), 'The impact of global warming on US agriculture: an econometric analysis of optimal growing conditions', *Review of Economics and Statistics*, **88**(1), 113–25.

Seo, S.N., R. Mendelsohn and M. Munasinghe (2005), 'Climate change and agriculture in Sri Lanka: a Ricardian valuation', *Environment and Development Economics*, **10**, 581–96.

5.   Modeling adaptation to climate change

Adaptations are actions that people and firms take in response to climate change to reduce damages or increase benefits (Carter et al. 1994; IPCC 2001; IPCC 2007b; Smith 1997; Smith and Lenhart, 1996). A comprehensive review of the adaptation literature can be found in Kurukulasuriya and Rosenthal (2003). There are three major strands to the adaptation literature: theoretical papers, on-site observations, and cross-sectional empirical studies. This chapter will provide a quick review of these strands and then present the modeling framework used in empirical studies of adaptation that are used in Chapter 9.

THEORY OF ADAPTATION

The theoretical literature stresses broad but important concepts. For example, efficient adaptations are actions that make actors better off (Fankhauser et al., 1999; Mendelsohn, 2000). Private adaptations are examples where there is only one beneficiary (the actor). Private adaptations will tend to be efficient because it is in the interest of the actor to make the change. As shown in Table 5.1, for example, farmers will switch crops or change planting dates to maximize profit. Public adaptations are examples where there are many beneficiaries. Building dams and canals or protecting endangered species are examples of public adaptations because there are many beneficiaries of these decisions. The market will have trouble making efficient public adaptations because the cost of coordination across multiple beneficiaries is large (especially if the beneficiaries are heterogeneous) (Mendelsohn, 2000, 2006). There is clearly a need for government assistance to champion efficient public adaptation.

Efficient adaptation relies on actors having private property rights (Mendelsohn, 2006). One serious problem in developing countries is the absence of these rights. For example, a great deal of land suitable for market uses is owned by the government. People living on this land would not necessarily adapt efficiently because they cannot get the benefits of protecting the land. There are similar problems with common property.

Table 5.1   Public and private agricultural adaptations to climate change

Type	Action
Private	Alter crop species and varieties
	Alter livestock species and breeds
	Alter timing of planting and harvesting
	Multiple cropping seasons
	Irrigation
	Alter land used for crops
	Alter land used for livestock and herds
Public	Plant and animal breeding
	Public education through extension
	Dams and canals

Source: Mendelsohn (2006).

Individual actors such as farmers, livestock owners and forest harvesters share in the benefits of land owned by local communities and so they underinvest in the natural capital. This leads to over-grazing, cropping on marginal lands, and deforestation. These problems with common property and public lands already exist but they will be exacerbated by climate change. People using these lands will not adapt efficiently.

Proactive adaptations are done in advance in anticipation of climate change. Proactive adaptations are important for long-lasting investments such as coastal defenses, dams, land use planning and roads. Actors must plan for how changes in climate affect these projects in the long run. However, it is difficult to forecast local climate change, so most adaptations will be reactive. The actor will wait until the climate change has occurred and then will act. As long as actions are relatively short-lived, most actors are better off relying on observations of recent weather to inform decisions. Using recent weather, reactive adaptation should be able to keep up with most climate changes. However, some adaptations are very long-lived. A dam may stay in place many decades, if not a century. Such long-lived investments must anticipate future climate because the future benefits will be affected.

The literature also stresses that timing matters (Mendelsohn, 2000; Sohngen et al., 2002; Yohe et al., 1999). Adaptations must fit the climate at every moment of time. If they are done too early, they tend to cost more and may have limited benefits. For example, if one plants a crop too early because it is expected to become warmer but it does not, the crop may well fail. If adaptations are done too late, climate damages mount and there are missed opportunities. For example, if one fails to build sea walls in time, rising sea levels will flood coastal property. Because climate change is dynamic, adaptation must also be dynamic, always trying to match adaptation to the climate at the time.

Some authors have argued that the best way to prepare for climate change is to adapt to climate variance, the changes in weather from year to year (Burton, 1997; Leary et al., 2006; Smit et al., 1996). They argue that adaptation is a stock. Building up that stock to address changes in weather, prepares the system to address changes in climate. For example, if farmers can choose crops and livestock that are productive in hot and dry weather, they will be prepared to make these choices to protect themselves against hot and dry climates. Further, these adaptations can begin today. It is not necessary to wait for climate change to occur.

There are circumstances where these arguments make sense. Weather will vary in the future just as it varies today (perhaps even more so; IPCC, 2007a). Coping with weather variation is likely to remain a problem for farmers. Learning to cope with interannual variations in weather makes sense. Some changes one would make to cope with annual weather may also resemble changes one would make if climate changes. For example, one may switch crops or change acreage if there is a forecast predicting a wetter or drier year. However, some adaptations that make sense for a year (the short term) do not make sense if the change is permanent. For example, selling off livestock in a bad year is not a long-run strategy for climate change although it may work well to smooth consumption against weather shocks. Insurance is a good policy for weather but not for climate change. If insurance compensates people for a bad crop year after year, they have no incentive to adapt. Changing capital and long-run investments makes sense for climate change but not for short-run weather shocks. There is no perfect parallel between adapting to weather and climate.

In addition to adaptation at the farm level, it is also true that markets adapt. For example, if climate change reduces the supply of a specific crop, the price of that crop will rise. This will induce farmers to plant more of that crop and moderate the initial reduction in supply. If climate change reduces crop yields per hectare, farmers may respond by planting more cropland (Lotsch, 2006; Mendelsohn et al., 1996). Similarly, international trade helps moderate the quantities of available crops in specific places (Reilly et al., 1994). For example, if climate change makes it difficult to grow wheat in Africa but easier to grow wheat in Russia, then Russian farmers can trade with African countries so that African consumers can still enjoy wheat at reasonable prices. International trade addresses some of the most severe problems that climate change will impose on production in local places, especially in the low latitudes (Reilly et al., 1994; Goklany,

1995). Of course, if global production falls, there will still be losses, but trade is likely to make those losses smaller.

Authors have also discussed the necessary conditions for adaptation (Leary et al., 2006; Mendelsohn, 2006). For example, is the scientific basis for adaptation established? Are people aware that climate has changed and are they aware of options they can take to adapt to it? Do people have adequate education and access to extension services? Is the institutional support available? Do farmers have access to capital markets and private property rights so that they can make long-term investments? Are regulations too stringent? Do farmers have the flexibility to make the changes they need to make?

OBSERVATIONAL CASE STUDIES

There is an extensive body of literature in environmental anthropology that reviews how people have adapted to weather and even to long-run shifts in regional climate. Most of these analyses are case studies of specific places where weather has changed and people have reacted. For example, there are studies of drought in semi-arid countries in the Middle East and in Africa. Analysts have done surveys and interviews to see how people coped with the change in weather. Farmers have often tried to adjust seeds and timing to obtain at least some crops in less fertile periods. They have also liquidated animal stocks in order to get through harsh drought conditions. Finally, they have sought income outside the farm. Some farmers work on commercial enterprises or on the farms of neighbors. Other farmers temporarily leave the community and seek work elsewhere (Adger et al., 2002).

In the long run, if local climate (long-run weather) actually shifts, there are more dramatic changes. Malthus (1798) records shifts in human populations in England along with shifts in weather patterns. There is growing evidence that the rise and fall of some ancient civilizations was concurrent with changes in local climate patterns (Weiss and Bradley, 2001; De Menocal, 2001).

EMPIRICAL ECONOMIC LITERATURE

This section reviews the economic literature concerning farm adaptation. This adaptation literature builds on earlier agricultural economics studies seeking to understand farm behavior. For example, earlier studies have looked at what determines irrigation choices (Caswell and Zilberman, 1986; Dinar and Yaron, 1990; Dinar and Zilberman, 1991; Dinar et al.,

1992; Negri and Brooks, 1990). These studies largely focused on economic factors that influence farmers' decisions, such as prices, income, access to capital, and extension services. This earlier literature captured farm adoption decisions but ignored climate adaptation.

Some of the first studies to look at adaptation were concerned that farmers might not adapt (Kaiser et al., 1993 and Kelly et al., 2005). They imagined that the decision making would be too difficult for farmers to grasp. Either the farmers would not know how to grow crops in the new climate or they would have trouble determining that the climate had changed. There would consequently be high adjustments costs as farmers went from one equilibrium to another. However, it is not clear that these authors are correct. Farmers quickly adapt to new prices and learn to grow different crops. Farm models already predict how farmers should shift from crop to crop as climate changes (Adams et al., 1999; Howitt and Pienaar, 2006). Further, farmers do not need a sophisticated climate projection to adapt in a timely fashion. They can simply rely on the average observed weather over the last five years to get a reasonable forecast of what weather will be like next year.

In addition to farm-level decisions, it is also clear that water decisions are important to agriculture. The water sector is often counted as a separate sector from agriculture, but the two sectors are closely linked. As water supply and as the demand for water changes with climate, there will be large changes in the amount of water available to farmers. It is important that water allocation changes in an efficient manner (Hurd et al., 1999; Lund et al., 2006). Water must be allocated across farmers efficiently. Water must also be shared with other users (urban and industrial) who may have higher marginal willingness to pay for water.

CROSS-SECTIONAL EMPIRICAL STUDIES

By comparing choices made by farmers who face different climate conditions, cross-sectional studies can uncover how farmers adapt to current climate. A researcher can examine numerous farm decisions using this approach including farm type, irrigation, livestock choice, crop choice, and a combination of them. In each case, the impact of climate on farmers' choices is tested. The sensitivity of these endogenous choices by farmers to climate reveals their climate adaptation. Presumably as climate changes, farmers will make these same adaptations to the new climate they face as they do now.

These cross-sectional methods were developed in several studies in many different locations. The method for studying farm type was first carried out for South America (Mendelsohn and Seo, 2007). The study of crop choice was carried out for both Africa (Kurukulasuriya and Mendelsohn, 2008b) and South America (Seo and Mendelsohn, 2008d). The analysis of livestock choice was carried out for Africa (Seo and Mendelsohn, 2008a, 2008b) and South America (Seo and Mendelsohn, 2007). Finally, the choice concerning irrigation was studied for Africa (Kurukulasuriya and Mendelsohn, 2008b) and South America (Mendelsohn and Seo, 2007).

This chapter demonstrates that multinomial choice models can capture the adaptations that farmers make to current climate. Multinomial logit models of adaptation can be estimated to capture how farmers adjust their endogenous choices in response to climate change. In each case, the model first estimates the choices that farmers make given their current climate. By looking at samples where farmers face widely different climates, the models can predict how climate affects each specific decision. These estimated models can then be used to predict how climate change would alter choices.

In order to estimate these choice models, we assume throughout that farmers maximize their net revenue per hectare. Net revenue is defined broadly to include the gross revenue consumed by the household and all costs except household labor. Farmers choose the desired farm type, irrigation technology, crop mix and livestock mix that yield the highest net revenue. Hence, the probability that a particular choice is made depends on the earnings from that choice. All of the choices follow the same theoretical model.

Farmer j's net revenue in making choice i from a set of choices j (j = 1, 2, . . ., N) is

$$\pi_{ij} = V_i(K_j, S_j) + \varepsilon_i(K_j, S_j) \tag{5.1}$$

where K is a vector of exogenous characteristics of the farm and S is a vector of characteristics of the farmer. The vector K includes climate, soils, prices and access variables. The household-specific vector S includes human capital variables, such as the age of the farmer, experience, and family composition and size. The net revenue function is composed of two components: the observable component V and an error term, ε. The error term is unknown to the analyst, but we assume it is known to the farmer. The farmer makes the choice that yields the highest net revenue. Defining $Z = (K, S)$, the farmer will choose i over all other choices if:

$$\pi_i^*(Z_j) > \pi_k^*(Z_j) \text{ for } \forall k \neq i.$$

$$[\text{or if } \varepsilon_k(Z_j) - \varepsilon_i(Z_j) < V(Z_j) - V_k(Z_j) \text{ for } k \neq i] \tag{5.2}$$

More succinctly, farmer j's problem is:

$$\arg\max_{i=1\ldots I_i}[\pi_1^*(Z_j), \pi_2^*(Z_j), \ldots, \pi_I^*(Z_j)] \tag{5.3}$$

The probability P_{ji} for choice i to be chosen by farmer j is then

$$P_{ji} = \Pr[\varepsilon_k(Z_j) - \varepsilon_i(Z_j) < V_i - V_k] \, \forall \, k \ne i \text{ where } V_i = V_i(Z_j) \tag{5.4}$$

Assuming ε is independently Gumbel distributed and $V_k = Z_{kj}\gamma_j + \alpha_k$,

$$P_{ji} = \frac{e^{Z_{ji}\gamma_j}}{\sum_{k=1}^{I} e^{Z_{jk}\gamma_j}} \tag{5.5}$$

which gives the probability that farmer (j) will choose (i) among (N) animals or crops (McFadden, 1981). This is the standard derivation of the multinomial logit model. The parameters of the model are γ_j.

One very basic choice facing farmers is farm type. Will they grow crops, livestock, or a combination of crops–livestock? If they choose to grow crops, they also have a choice of using rainfed systems or irrigation (if water is available). One way to model the choice of farm type is to envision five basic choices: rainfed crops, irrigated crops, livestock, rainfed crop–livestock combination, and irrigated crop–livestock combination. The farmer must choose from these five alternatives the one that maximizes net revenue.

An alternative strategy is to assume a nested structure where the farmer first chooses crops, livestock, or the combination of crops–livestock. Conditional on growing crops, the farmer then chooses whether or not to irrigate. A variant of this nested model would examine the conditional choice of irrigation, depending on whether the farm grows just crops or whether the farmer chooses the crop–livestock combination. This second variation examines the possibility that farmers who select crops alone may behave differently from farmers who choose the crop–livestock combination. The different modeling approaches imply slightly different assumptions about the independence of irrelevant alternatives.

Farmers must also choose which species of crop to grow or livestock to raise. Although it is well known that particular crops or particular animals are associated with current regions and therefore current climates, the climate change literature has been slow to grasp that these choices made by the farmer will be affected by climate change. The agronomic studies of climate change in particular tend to assume that crop choice is attached to the geography rather than to the climate (for example Rosenzweig and

Parry, 1994; Rosenzweig and Hillel, 1998). Whether crop choices or live-stock choices are actually a function of climate can then be tested.

One intriguing complication associated with livestock and crop choice is that farmers can choose more than one species at a time. There are many combinations of animals that the farmer could choose. There are conse-quently three ways that the choice can be modeled (Seo and Mendelsohn, 2007b). One can model the primary animal, the portfolio of animals, or the demand for each species.

The primary crop or animal model is defined as the single crop or animal that earns the most revenue on the farm. The primary crop or animal model focuses solely on which primary animal or crop each farmer chooses. Although this approach ignores secondary species, the primary crop or animal can sometimes dominate the revenues of the farm. Farmers may have a low valued second season with a secondary crop but the primary crop may earn the principal net revenue. Similarly, some farmers may have secondary animals for household purposes but the main commercial animal is the primary animal. The primary crop or animal model has the advantage that there is a single choice and it is mutually exclusive.

Another alternative is to model the portfolio of crops (crop mix) and animals chosen by each farmer. Instead of focusing on one animal, this approach treats combinations of animals or crops as a choice. For example, a farmer might grow maize and beans as a choice while another farmer might raise goats and chickens. The portfolio approach captures the full complexity of farm choices but at a cost. As the number of species, n, increases, the number of potential combinations increases by 2^{n-1}. Further, different combinations of choices can be close substitutes for one another. For example, a combination of maize and millet may behave quite similarly to millet alone. Under such situations, the modeling can quickly become too complex.

The portfolio approach, however, can capture one important phe-nomenon that the other approaches cannot. The portfolio approach can reflect how farmers diversify their choices to cope with climate variation. The more year to year weather variation a farmer faces, the more differ-ent crops he/she might want to plant or different livestock he/she might want to own. The diversification of crops or livestock provides protection against uncertainty.

A third approach that can be used to capture species choice is a system of demand functions (Seo and Mendelsohn 2007b). The farmer can select whether to choose each crop or animal independently from the other choices. Because the choices in this model are not mutually exclusive, the totals do not need to add to one.

The demand system can be estimated as a system of equations using multivariate probit (Chib and Greenberg, 1998). Let Y_{ij} denote the binary response of ith farmer on the jth animal or crop and let $Y_i = (Y_{i1}, \ldots, Y_{iJ})$ denote the collection of responses on all J animals or crops. According to the multivariate probit model, the probability that $Y_i = y_i$, conditioned on parameters β, Σ, and a set of covariates x_{ij}, is given by

$$P(Y_i = y_i | \beta, \Sigma) = \int_{A_{iJ}} \ldots \int_{A_{i1}} \phi_J(t_1, \ldots, t_J | 0, \Sigma) dt_1 \ldots dt_J \qquad (5.6)$$

where $\phi_J(t|0, \Sigma)$ is J-variate normal distribution with mean vector 0 and correlation matrix $\Sigma = \{\sigma_{jk}\}$, and A_{ij} is the interval within which selection decisions are made.

$$A_{ij} = \begin{cases} (-\infty, x'_{ij}\beta_j) & \text{if } y_{ij} = 1, \\ (x'_{ij}\beta_j, \infty) & \text{if } y_{ij} = 0. \end{cases} \qquad (5.7)$$

Given the estimated models described above, we now turn to examining the influence of climate change on farmers' adaptation decisions. We assume that farmers will adjust to future climates just as they have adjusted to current ones. That is, they will use the existing technologies designed for each climate zone. The estimated model of farmer choice can then be used to predict adaptation to future climate change. For example, suppose that farmer j currently has the following probability of making each choice i depending on an initial climate C_{0j} such as $P_i(C_{0j}, Z_j)$ where Z_j is a set of relevant factors specific to the farm. Given the empirical model explaining how climate affects choice i, we can postulate that if climate changes to C_{1j}, farmer j would change his/her probability of choosing i to:

$$\Delta P_{ij} = P_i(C_{1j}, Z_j) - P_i(C_{0j}, Z_j). \qquad (5.8)$$

The empirical model predicts how climate change would affect the choices of each farmer j. These changes can then be mapped to see how they change across the landscape. The changes can also be aggregated to see what difference they make to aggregate outcomes.

REFERENCES

Adams, Richard, Bruce McCarl, Kathy Segerson, Cynthia Rosenzweig, Kelley Bryant, Bruce Dixon, Richard Conner, Robert Evenson and Dennis Ojima (1999), 'The economic effects of climate change on US agriculture', in

R. Mendelsohn and J. Neumann (eds), *The Impact of Climate Change on the United States Economy*, Cambridge, UK: Cambridge University Press, pp. 18–54.

Adger, W., P. Kelly, A. Wikels, L. Huy and C. Locke (2002), 'Migration, remittances, livelihood trajectories and social resilience', *Ambio*, **31**, 358–66.

Burton, I. (1997), 'Vulnerability and adaptive response in the context of climate and climate change', *Climatic Change*, **36**, 185–96.

Carter, T., M. Parry, H. Harasawa and S. Nishioka (1994), *IPCC Technical Guidelines for Assessing Climate Change Impacts and Adaptations*, London: Department of Geography, University College, London.

Caswell, Margiarette and David Zilberman (1986), 'The effect of well depth and land quality on the choice of irrigation technology', *American Journal of Agricultural Economics*, **68**, 798–811.

Chib, S. and E. Greenberg (1998), 'An analysis of multivariate probit models', *Biometrika*, **85**, 347–61.

De Menocal, P. (2001), 'Cultural responses to climate change during the late holocene', *Science*, **292**, 667–73.

Dinar, Ariel and Dan Yaron (1990), 'Influence of quality and scarcity of inputs on the adoption of modern irrigation technologies', *Western Journal of Agricultural Economics*, **15**(2), 224–33.

Dinar, Ariel and David Zilberman (1991), 'The economics of resource-conservation, pollution–reduction technology selection, the case of irrigation water', *Resources and Energy*, **13**, 323–48.

Dinar, Ariel, Mark B. Campbell and David Zilberman (1992), 'Adoption of improved irrigation and drainage reduction technologies under limiting environmental conditions', *Environmental & Resource Economics*, **2**(4), 373–98.

Dubin, Jeffrey A. and Daniel L. McFadden (1984), 'An econometric analysis of residential electric appliance holdings and consumption', *Econometrica*, **52**, 345–62.

Fankhauser, S., J. Smith and R. Tol (1999), 'Weathering climate change: some simple rules to guide adaptation decisions', *Ecological Economics*, **30**, 67–78.

Goklany, I. (1995), 'Strategies to enhance adaptability: technological change, sustainable growth, and free trade', *Climatic Change*, **30**, 427–49.

Greene, William H (1998), *Econometric Analysis*, 3rd edn, New Jersey, USA: Prentice Hall.

Howitt, R. and E. Pienaar (2006), 'Agricultural impacts', in J. Smith and R. Mendelsohn (eds), *The Impact of Climate Change on Regional Systems: A Comprehensive Analysis of California*, Cheltenham, UK and Northampton, MA, USA: Edward Elgar, pp. 188–207.

Hurd, B., J. Callaway, J. Smith and P. Kirshen (1999), 'Economics effects of climate change on US water resources', in R. Mendelsohn and J. Neumann (eds), *The Impact of Climate Change on the United States Economy*, Cambridge, UK: Cambridge University Press, pp. 133–77.

IPCC (2001), *Climate Change 2001: Impacts, Adaptation and Vulnerability*, contribution of Working Group II to the Third Assessment Report of the Intergovernmental Panel on Climate Change, Cambridge, UK: Cambridge University Press.

IPCC (2007a), *Climate Change 2007: The Physical Science Basis*, contribution of Working Group I to the Fourth Assessment Report of the Intergovernmental Panel on Climate Change, Cambridge, UK: Cambridge University Press.

IPCC (2007b), *Impacts, Adaptation and Vulnerability*, contribution of Working

Group II to the Fourth Assessment Report of the Intergovernmental Panel on Climate Change, Cambridge, UK: Cambridge University Press.

Kaiser, H.M., S.J. Riha, D.S. Wilkes, D.G. Rossiter and R.K. Sampath (1993), 'A farm-level analysis of economic and agronomic impacts of gradual warming', *American Journal of Agricultural Economics*, **75**, 387–98.

Kelly, D.L., C.D. Kolstad and G.T. Mitchell (2005), 'Adjustment costs from environmental change', *Journal of Environmental Economics and Management*, **50**(3), 468–95.

Kurukulasuriya, P. and R. Mendelsohn (2008a), 'Modeling endogenous irrigation: the impact of climate change on farmers in Africa', World Bank Policy Research Working Paper 4278, Washington, DC, USA: World Bank.

Kurukulasuriya, P. and R. Mendelsohn (2008b), 'Crop switching as an adaptation strategy to climate change', *African Journal of Agriculture and Resource Economics*, **2**, 1–23.

Kurukulasuriya, P. and S. Rosenthal (2003), 'Climate change and agriculture: a review of impacts and adaptations', Environment Department Paper No. 91, Washington, DC, USA: World Bank.

Leary, N., M.W. Baethgen, V. Barros, I. Burton, O. Canziani, T.E. Downing, R. Klein, D. Malpede, J.A. Marengo, L.O. Mearns, R.D. Lasco, and S.O. Wandiga (2006), 'A plan of action to support climate change adaptation through scientific capacity, knowledge and research', AIACC Report 23, International START Secretariat, Washington, DC, USA.

Lotsch, A. (2006), 'Sensitivity of cropping patterns in Africa to transient climate change', CEEPA Discussion Paper No. 14, University of Pretoria, South Africa, available at www.ceepa.co.za/Climate_Change/project.html.

Lund, J., T. Zhu, S. Tanaka and M. Jenkins (2006), 'Water resource impacts', in J. Smith and R. Mendelsohn (eds), *The Impact of Climate Change on Regional Systems: A Comprehensive Analysis of California*, Cheltenham, UK and Northampton, MA, USA: Edward Elgar, pp. 165–87.

Malthus, Thomas (1798), *An Essay on the Principle of Population*, London, UK: J. Johnson.

McFadden, Daniel L. (1981), 'Econometric models of probabilistic choice', in C.F. Manski and D.L. McFadden (eds), *Structural Analysis of Discrete Data and Econometric Applications*, Cambridge, MA, USA: MIT Press.

Mendelsohn, R. (2000), 'Efficient adaptation to climate change', *Climatic Change*, **45**, 583–600.

Mendelsohn, R. (2006), 'The role of markets and governments in helping society adapt to a changing climate', *Climatic Change*, Special Issue on Climate, Economy, and Society: from Adaptation to Adaptive Management, **78**, 203–15.

Mendelsohn, R. and S.N. Seo (2007), 'An integrated farm model of crops and livestock: modeling Latin American agricultural impacts and adaptation to climate change', World Bank Policy Research Working Paper 4161, Washington, DC, USA: World Bank.

Mendelsohn, R., W. Nordhaus and D. Shaw (1996), 'Climate impacts on aggregate farm values: accounting for adaptation', *Agriculture and Forest Meteorology*, **80**, 55–67.

Mendelsohn, R., A.F. Avila and S.N. Seo (2007), *Synthesis of the Latin American Results*, Montevideo, Uruguay: PROCISUR.

Negri, Donald H. and Douglas H. Brooks (1990), 'Determinants of irrigation technology choice', *Western Journal of Agricultural Economics*, **15**(2), 213–23.

Pearce, D. et al (1996), 'The social costs of climate change: greenhouse damage and benefits of control', in IPCC, *Climate Change 1995: Economic and Social Dimensions of Climate Change*, Cambridge, UK: Cambridge University Press, pp.179–224.

Reilly, J., N. Hohnmann and S. Kane (1994), 'Climate change and agricultural trade: who benefits and who loses?', *Global Environmental Change*, **4**, 24–36.

Rosenzweig, C. and M.L. Parry (1994), 'Potential impact of climate change on world food supply', *Nature*, **367**, 133–8.

Rosenzweig, C. and D. Hillel (1998), *Climate Change and the Global Harvest: Potential Impacts of the Greenhouse Effect on Agriculture*, New York, USA: Oxford University Press.

Schneider, S., W. Easterling, L.O. Mearns (2000), 'Adaptation: sensitivity to natural variability, agent assumptions, and dynamic climate changes', *Climatic Change*, **45**, 203–21.

Seo, S.N. and R. Mendelsohn (2007a), 'An analysis of livestock choice: adapting to climate change in Latin American farms', World Bank Policy Research Working Paper 4164, Washington, DC, USA: World Bank.

Seo, N.S. and R. Mendelsohn (2007b), 'Climate change adaptation in Africa: a microeconomic analysis of livestock choice', World Bank Policy Research Working Paper 4277, Washington, DC: World Bank.

Seo, S.N. and R. Mendelsohn (2008a), 'Measuring impacts and adaptation to climate change: a structural Ricardian model of African livestock management', *Agricultural Economics*, **38**, 150–65.

Seo, S.N. and R. Mendelsohn (2008b), 'Climate change impacts and adaptations on animal husbandry in Africa', *African Journal of Agriculture and Resource Economics*, **2**, 65–82.

Seo, S.N. and R. Mendelsohn (2008c), 'A Ricardian analysis of the impact of climate change on South American farms', *Chilean Journal of Agricultural Research*, **68**(1), 69–79.

Seo, S.N. and R. Mendelsohn (2008d), 'An analysis of crop choice: adapting to climate change in Latin American farms', *Ecological Economics*, **67**, 109–16.

Smit, B., D. McNabb and J. Smithers (1996), 'Agricultural adaptations to climate variation', *Climatic Change*, **33**, 7–29.

Smith, J.B. (1997), 'Setting priorities for adaptation to climate change', *Global Environmental* Change, **7**, 251–64.

Smith, J.B. and S.S. Lenhart (1996), 'Climate change adaptation policy options', *Climate Research*, **6**, 193–201.

Sohngen, B., R. Mendelsohn and R. Sedjo (2002), 'A global model of climate change impacts on timber markets', *Journal of Agricultural and Resource Economics*, **26**, 326–43.

Tol, R., S. Fankhauser and J. Smith (1998), 'The scope for adaptation to climate change: what can we learn from the impact literature?', *Global Environmental Change*, **8**, 109–23.

Weiss, H. and R.S. Bradley (2001), 'What drives societal collapse?', *Science*, **291**, 609–10.

Yohe, G., J. Neumann and P. Marshall (1999), 'The economic damage induced by sea level rise in the United States', in R. Mendelsohn and J. Neumann (eds), *The Impact of Climate Change on the United States Economy*, Cambridge, UK: Cambridge University Press, pp. 178–208.

6. Structural Ricardian models

This chapter blends the insight of the traditional Ricardian model (Chapter 4) and the explicit adaptation model (Chapter 5) into a powerful tool that captures both changes in expected income from and adaptation to climate change. The combined tool is named a 'Structural Ricardian model' (Seo and Mendelsohn, 2008). This multiple stage model first estimates farm choices and then estimates the conditional income for each choice. The model uses cross-sectional evidence to measure not only the adaptive choices made by farmers but also how climate change affects expected income.

As stated in Chapter 4, the traditional Ricardian approach can estimate the expected effect of climate on farm land value or net revenue (Mendelsohn et al., 1994; Kurukulasuriya et al., 2006). However, the traditional Ricardian model does not provide insight into how farmers adapt to climate because adaptation is treated endogenously. The reduced form analysis treats farm decision as a 'black box' that is not visible to the analyst. In order to learn about adaptation, one must get beneath the Ricardian locus of profit maximizing combinations of choices and begin to distinguish each choice by a farmer and the consequences of each choice on profit. The Structural Ricardian model is attempting to estimate the underlying choice model that leads to the traditional Ricardian locus.

One suggestion in the spirit of the Structural Ricardian model is to estimate a separate Ricardian regression for rainfed and for irrigated land (Schlenker et al., 2005). The regression would distinguish the climate sensitivity of these two farming alternatives. However, it is not enough just to run separate Ricardian regressions because it implies the two choices are exogenously determined, which is not the case. Further, the analysis is vulnerable to sample selection bias.

In order to capture adaptation within the Ricardian framework, it is necessary to first explicitly model the adaptation choice such as farm type, irrigation, crop choice, or livestock choice. Given the selection, one can then estimate conditional Ricardian models for each choice. Because this leads to a sample selection problem (Heckman, 1979), it is important to control for the correlation in errors between the choice and income equations.

The next section of the chapter develops a theoretical model of the farm with farmers first choosing a farm type, irrigation, crop species, or livestock species based on climate and other exogenous conditions (Kurukulasuriya and Mendelsohn, 2008a, 2008b; Mendelsohn and Seo, 2007; Seo and Mendelsohn, 2008). The choice is estimated using a probit, logit or multinomial logit. Given the choice of farm type, irrigation, crop species, or livestock species, a second stage estimates the conditional net revenue for each choice. We illustrate this general model using the choice of farm type. The third section describes how to use the estimated model to forecast impacts and adaptations across a set of future climate scenarios. One can estimate the value of adaptation by comparing the change in income with and without the adaptation. We conclude the chapter discussing the policy implications and the limitations of the structural Ricardian approach.

THEORETICAL MODEL

We assume that farmers make a choice amongst a set of alternatives to maximize the net revenue or value of the land. For example, the farmer may want to choose amongst the following types of farms: only rainfed crops, only irrigated crops, rainfed crops and livestock (mixed), irrigated crops and livestock (mixed), and only livestock farming. Alternatively, the farmer may want to choose what crop to grow or what type of livestock to own. Given these choices, the farmer combines inputs to make outputs that maximize land value. We assume that the farmer will choose the combination of farm type, irrigation, and species that maximizes expected net revenues. The first stage of the Structural Ricardian model is the same as the model used to study adaptation in Chapter 5.

In Figure 6.1, we show a hypothetical relationship between farm type and climate. The picture suggests that each farm type is ideal for a particular climate range. As climate changes, farmers switch from one farm type to another. The overall response function, the envelope, captures this switching. The Structural Ricardian model explicitly captures the switching as farmers stay on the maximum profit locus at different temperatures or precipitation levels.

More formally, each farmer j maximizes net revenue by making a choice i ($i=1, \ldots, I$):

$$\pi_i = X_j\beta_i + u_i \tag{6.1}$$

$$\pi_{ij}^* = Z_j\gamma_i + \eta_i, \quad i = 1, \ldots, I. \tag{6.2}$$

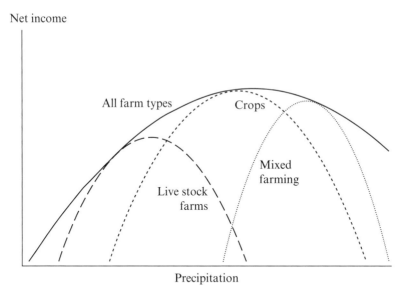

Source: Adapted from Dinar et al. (2008, Figure 3.1).

Figure 6.1   Multiple conditional income functions

where $E(u_i|X, Z) = 0$ and $\text{var}(u_i|X, Z) = \sigma^2$. The subscript i is a categorical variable indicating the choice amongst I alternatives. The vector Z represents the set of explanatory variables that determine the choice amongst the alternatives. The vector X contains the determinants of income given the choice. Following the same steps as outlined in Chapter 5, the probability P_i of farm type i being chosen is:

$$P_{ij} = \frac{\exp(Z_j\gamma_i)}{\sum_{k=1}^{K}\exp(Z_j\gamma_k)}. \tag{6.3}$$

The choice equation is identified using cross price terms and other identifying variables.

Given the choice of i, the farmer will choose inputs and outputs to maximize a conditional net revenue function from observations that selected choice i. This second equation involves a regression of net revenue or conditional land value on a set of exogenous variables X. However, it is likely that the errors in equation (6.1), the selection equation, and equation (6.2), the conditional income equation, are correlated because farmers may know information unknown to the researcher. The farmers will tend to select the choice that yields the best outcome for them (Heckman, 1979). As profits for choice i are only observed for the farms that choose farm type i, selection bias should be corrected to obtain consistent estimates of the parameters (Heckman, 1979). Following Dubin and McFadden (1984), we assume the following linearity condition:

$$E(u_1|\eta_1,...,\eta_J) = \sigma \cdot \sum_{j=1}^{J} r_j \cdot (\eta_j - E(\eta_j)) , \text{ with } \sum_{j=1}^{J} r_j = 0 \quad (6.4)$$

We can estimate the conditional profit function for choice i as follows:

$$\pi_i = X_i \varphi_i + \sigma \cdot \sum_{k \neq 1}^{K} r_k \cdot \left[\frac{P_k \cdot \ln P_k}{1 - P_k} + \ln P_i \right] + \delta_i \quad (6.5)$$

Note that η in equation (6.2) and δ in equation (6.5) are now independent.

The regressors in the above equation include soils, climate and socio-economic variables. Country dummy variables are also tested to see if country-specific conditions (country fixed effects) make a substantial difference. We follow the previous literature concerning climate and assume seasons continue to matter and that climate effects have a quadratic shape.

As discussed in Chapter 4, one can employ land value or net revenue as the measure of net productivity of land for conditional income. With perfect competition for land, free entry and exit will drive excess profits to zero on the margin. In this case, land rents will equal net revenue per hectare. Land value will reflect the present value of the net revenue of each farm.

The expected value of the farm, V, is the sum of the probabilities across each choice the farmer faces, multiplied by the conditional land value of that farm type. That is:

$$V_i(C) = \sum_{k=1}^{K} P_k(C_i, Z_i) \cdot \pi_k(C_i, Z_i) \quad (6.6)$$

Note that this measure does not assume that choices remain constant.

CLIMATE CHANGE IMPACTS SIMULATIONS

In this section, we discuss how to use the Structural Ricardian model to predict the impact of future climate scenarios. As discussed in earlier chapters, there are several caveats one must keep in mind with such forecasts.

One must check whether there are omitted variables that have biased the results; one must also check whether other factors have changed, such as technology, prices and carbon dioxide.

The change in welfare, ΔW, resulting from a climate change from C_0 to C_1 can be measured as follows:

$$\Delta W = [V(C_1, Z_i) - V(C_0, Z_i)] \cdot L_i$$

$$= \sum_{k=1}^{K} \left[P_k(C_1) \cdot \pi_k(C_1) - \sum_{k=1}^{K} P_k(C_0) \cdot \pi_k(C_0) \right] \cdot L_i \qquad (6.7)$$

where L_i is the amount of land at each farm i. Note that this welfare calculation for the Structural Ricardian model is quite similar to the welfare calculation for the traditional Ricardian model (equation (4.7) in Chapter 4). The difference is that the traditional Ricardian welfare calculation depends upon a single equation measuring the overarching locus whereas the Structural Ricardian welfare calculation depends on the combination of the choice and conditional income results. The change in welfare captures two effects. Climate is likely to change the probability of each choice. Second, climate will affect the conditional income of each choice. Both of these changes are captured in the expected outcome measure.

In order to test the importance of adaptation, one can compare the expected income of climate change without adaptation with the expected income of climate change with adaptation. The initial expected income of the farmer at the current climate, C_0, is the same whether one models adaptation or not because everyone has adapted to the current climate. Adaptation only matters in this model if climate changes. With adaptation, the farmers change their set of choices to raise their expected income with the new climate. Without adaptation, the farmers keep the same set of choices made under the old climate:

$$W(C) = \sum_{k=1}^{K} \left[P_k(C_1) \cdot \pi_k(C_0) - \sum_{k=1}^{K} P_k(C_0) \cdot \pi_k(C_0) \right] \cdot L_i \quad (6.8)$$

Without adaptation, farmers will continue to make the same choices that they make now. As the conditional incomes of each choice change, this will make farmers worse off than if they adapted. The difference in welfare between equations (6.7) and (6.8) measures the value of adaptation.

This chapter blends two strands of the climate change economic literature to develop a more in-depth understanding of the effect of climate change on agriculture. One strand is the adaptation literature, which explains how climate influences farmers' decisions. The other strand is the Ricardian literature, which has developed an effective method of measuring climate change impacts across the world. By marrying the two approaches, the Structural Ricardian model reveals the precise adaptations that farmers are making to climate change, the value of these adaptations, and the net economic impact of climate change. By revealing more of the details about what is actually happening on farms, this new approach moves from the black box nature of the original Ricardian model towards the detailed portrayal of farms in the agronomic–economic literature.

The chapter explicitly discusses the choice of farm type, irrigation, crop choice and livestock choice. However, many other choices made by farmers could also be captured with the above methods, including the number of seasons, machinery purchases, and types of irrigation technology. Continuous choices such as tons of fertilizer or pesticides or length of growing season could be modeled as well in the first stage choice model. The model of the farm could also be made more complicated by nesting stages of choices with a final stage of conditional income equations.

REFERENCES

Dubin, Jeffrey A. and Daniel L. McFadden (1984), 'An econometric analysis of residential electric appliance holdings and consumption', *Econometrica*, **52**(2), 345–62.

Heckman, James J. (1979), 'Sample selection bias as a specification error', *Econometrica*, **47**, 153–62.

Kurukulasuriya, P. and R. Mendelsohn (2008a), 'Crop switching as an adaptation strategy to climate change', *African Journal of Agriculture and Resource Economics*, **2**, 105–26.

Kurukulasuriya, P. and R. Mendelsohn (2008b), 'Modeling endogenous irrigation: the impact of climate change on farmers in Africa', World Bank Policy Research Working Paper 4278, Washington, DC, USA: World Bank.

Kurukulasuriya, P., R. Mendelsohn, R. Hassan, J. Benhin, M. Diop, H.M. Eid, K.Y. Fosu, G. Gbetibouo, S. Jain, A. Mahamadou, S. El-Marsafawy, S. Ouda, M. Ouedraogo, I. Sène, S.N. Seo, D. Maddison and A. Dinar (2006), 'Will African agriculture survive climate change?', *World Bank Economic Review*, **20**, 367–88.

Mendelsohn, R. and A. Dinar (2003), 'Climate, water, and agriculture', *Land Economics*, **79**, 328–41.

Mendelsohn, R. and S.N. Seo (2007), 'Changing farm types and irrigation as an adaptation to climate change', World Bank Policy Research Working Paper 4161, Washington, DC, USA: World Bank.

Mendelsohn, R., W. Nordhaus and D. Shaw (1994), 'The impact of global warming on agriculture: a Ricardian analysis', *American Economic Review*, **84**, 753–71.

Schlenker, Wolfram, Michael Hanemann and Anthony Fisher (2005), 'Will US agriculture really benefit from global warming? Accounting for irrigation in the hedonic approach', *American Economic Review*, **95**, 395–406.

Seo, S.N., R. Mendelsohn and M. Munasinghe (2005), 'Climate change and agriculture in Sri Lanka: a Ricardian valuation', *Environment and Development Economics*, **10**, 581–96.

Seo, S.N. and R. Mendelsohn (2008), 'Measuring impacts and adaptation to climate change: a structural Ricardian model of African livestock management', *Agricultural Economics*, **38**, 150–65.

7. Ricardian analyses of aggregate data

This chapter reviews various Ricardian studies that have relied on aggregate data that reports average responses across all farmers from a county, district or municipio. The aggregate data is attractive because it is collected by agricultural census administrations in various countries, is pre-existing, and accessible. Unfortunately, such data is not available in many countries and it is often not consistent across countries. In order to use this data, one generally needs a single country with good agriculture census capabilities that is also large enough for one to observe farms across different climate zones. It is not an accident that the first Ricardian studies were done in the US, India and Brazil, countries known for their consistent and reliable census data. However, in some circumstances, it is possible to study even a small country such as Sri Lanka or Israel if there is enough variation in desired attributes, including climate.

As discussed in Chapters 3 and 4, the Ricardian method is a cross-sectional method that regresses land value or net revenue on climate, soils and other control variables. Because it is based on climate and not weather, the method is examining the equilibrium response of each farm to the climate it is in. The equilibrium response captures secondary effects such as differences in diseases, pests and ecosystems that are caused by climates. The equilibrium response also captures the long-run adaptations of the farmer since each farmer has adapted to where they live.

US STUDIES

The first applications of the Ricardian method relied on county data from the United States (Mendelsohn et al., 1994, 1996, 1999). Climate was measured using linear and squared terms for temperature and precipitation. The seasonal component was reflected by using monthly normals for January, April, July and October. A normal is the mean weather over a 30-year period, in this case, between 1960 and 1990.

One practical problem facing all cross-sectional climate studies is obtaining accurate climate data for each location. Weather stations are not necessarily available for each location. In the initial Ricardian studies in the US, monthly normals for counties were interpolated from weather stations within 500 miles of the county. Weighting each station by 1/ distance, a weather surface for each monthly normal was estimated using the weather stations within 500 miles of each county. The predicted value of climate for the county was taken from that surface. A separate surface was estimated for each county and each month.

The farm data used in the Ricardian regressions was from the US Census of Agriculture for 1978 and 1982. Economic data came from the Census of Housing and Population (US Census Bureau, 1988). Soils data came from the National Resource Inventory of the US Department of Agriculture. Soil statistics were calculated from soil plots in each county.

The dependent variable in all the US regressions was farmland value per hectare. Farmland value includes the value of land and buildings. The independent variables in the regression included linear and quadratic seasonal climate variables, soils, and other control variables such as elevation, population density and farm size. Two weights were utilized to place more emphasis on counties that grow crops. The percentage of the county in cropland placed more emphasis on regions of the country where more land is devoted to crops. In practice, this places more emphasis on the grain-growing counties of the Great Plains. Gross crop revenue was also used to reflect counties making a large contribution to crop GDP. In addition to reflecting counties growing grains, this measure also emphasized counties growing fruits and vegetables (irrigated lands in the southern tier of the US). There was no attempt in this first study to measure the impact of climate on livestock.

The results of the regressions with all the explanatory variables are shown in Table 7.1. The regressions explain a great deal of the spatial variation in the data (the R squared varies from 78 percent to 84 percent). The climate variables in every season are significant. Although it was not surprising that climate during the growing season is important, the analysis reveals that winter is also important. Winter has a role to play in removing pests and in providing soil moisture for the start of the growing season. The squared terms are significant, implying that the shapes of the seasonal climate response functions are non-linear. Agronomic results suggest that the squared temperature terms should be negative. This proved true in January, April and July. However, there was a large positive coefficient for the squared temperature term in October. Some critics felt that this positive squared term implies the regression is misspecified (Quiggin and Horowitz, 1999). However, warm temperatures in autumn help ripen fruits and grains and so the positive effect in this season is plausible. For example, some farmers resort to applying heat in order to encourage certain crops to ripen.

Table 7.1   US Ricardian regression

Variable	1982 Crop land	1978 Crop land	1982 Crop revenue	1978 Crop revenue
Constant	1329**	1173**	1451**	1307**
Jan. temp.	−88.6**	−103**	−160**	−138**
Jan. temp. sq.	−1.34**	−2.11**	−2.68**	−3.00**
Apr. temp.	−18.0	23.6*	13.6	31.8*
Apr. temp. sq.	−4.90**	−4.31**	−6.69**	−6.63**
Jul. temp.	−155**	−177**	−87.7**	−132**
Jul. temp. sq.	−2.95**	−3.87**	−0.30	−1.27*
Oct. temp.	192**	175**	217**	198**
Oct. temp. sq.	6.62**	7.65**	12.4**	12.4**
Jan. prec.	85.0**	56.5*	280**	172**
Jan. prec. sq.	2.73	2.20	−10.80*	−4.09
Apr. prec.	104**	128**	82.8*	113**
Apr. prec. sq.	−16.5*	−10.8	−62.1**	−30.6*
Jul. prec.	−34.5*	−11.3	−116**	−5.3
Jul. prec. sq.	52.0**	37.8**	57.0**	34.8**
Oct. prec.	−50.3*	−91.6**	−124.0*	−135.0**
Oct. prec. sq.	2.3	0.3	171.0**	106.0**
Income	71.0**	65.3**	48.5**	47.1**
Density	130**	105**	153**	117**
Density sq.	−1.72**	−0.93*	−2.04**	−0.94**
Latitude	−90.5**	−94.4**	−105.0**	−85.8**
Altitude	−167**	−161**	−163**	−149**
Salinity	−684*	−416*	−582*	−153
Flooding	−163*	−309**	−663**	−740**
Wetland	−58.2	−57.5	762**	230
Soil erosion	−1258**	−1513**	−2690**	−2944**
Slope length	17.3*	13.7*	54.0**	30.9**
Sand	−139*	−36	−288**	−213**
Clay	86.2**	67.3*	−7.9	−18.0
Moisture	37.7**	51.0**	20.6*	45.0**
Permeability	−0.2	−0.5*	−1.3**	−1.7**
Adjusted R^2	0.782	0.784	0.836	0.835

Notes:
Dependent variable is farmland value in USD per acre.
Coefficients significant at 5% level are marked with one asterisk (*) and at 1% level with two asterisks (**).

Source: Mendelsohn et al. (1994, Table 3).

The control variables in Table 7.1 have the expected signs. Income, population density and soil moisture capacity have the expected positive effect on land values. Latitude, altitude, salinity, flooding, wetland, soil erosion, sand and permeability all have negative effects as expected. Slope length has a positive effect. Slope length is the distance before the slope changes and it basically measures how flat the land is. Longer slopes are easier for machines to manage.

In order to get a sense of how climate affects current land values, we present two maps of climate effects. Figure 7.1 (see p. 122) presents a map using cropland weights for each county. The map explains how much of the variation in farmland values can be explained by climate alone given the coefficients in Table 7.1. Note that the southern tier of the US has much lower values because of the warm temperatures in that region. The central wedge of the country has average climate effects but the northern and northwestern regions benefit from their existing climate relative to the rest of the country. Figure 7.2 (see p. 123) presents the same map except that counties are weighted by their contribution to crop revenue instead of cropland. One still sees large negative effects in the southern tier of the US but they do not extend into California. It is likely that the harmful effects of climate are offset by surface irrigation water in this region, leading to a very different perspective. Another difference is that with crop revenue weights, the climates of the Rocky Mountains appear to be more beneficial.

In order to understand how future climate effects may be distributed across space, we present two maps of future climate simulation results. Figure 7.3 (see p. 124) examines the mild warming scenario predicted by the PCM climate model for 2100 (Washington et al., 2000). The results reveal that the entire eastern US would benefit in this scenario. The only regions that would be damaged are the northern Rocky Mountains and especially the Pacific coast from Washington to California. Figure 7.4 (see p. 125) presents the US results for the more extreme HadCM3 scenario for 2100 (Hadley Centre, 2008). Figure 7.4 is different from Figure 7.3. The large gains in the midwest in Figure 7.3 have shrunk noticeably in Figure 7.4. The western plains are starting to show damages. The damages in California have expanded from the coast. There is no question that the Hadley scenario leads to smaller benefits than the PCM scenario.

Subsequent research in the US includes an analysis of cropland per acre of land (Mendelsohn et al., 1996). Instead of using farmland value per hectare of farmland as the dependent variable, this analysis uses farmland value per hectare of total land in a county. The dependent variable is the product of farmland value per hectare of farmland times farmland hectares divided by total land. The analysis not only captures the change in land value but also the change in farmland. The temperature and precipitation results are quite similar to the results of the first analysis. However, one difference is that the sum of the squared terms for temperature is negative as expected. The total value of farmland has a hill-shaped relationship with respect to temperature.

Two additional analyses investigate the importance of interannual variance (Mendelsohn et al., 1999, 2007b). Both analyses reveal that land values are affected by interannual variance. Higher variance generally leads to lower values. An exception to this rule is winter variance. The result here suggests that higher variance is beneficial. Farmers can observe winter weather before they commit to planting and so they can fully adapt to winter variance. In the other seasons, they are already committed before the weather in that season is known. So there are only limited and relatively costly adaptation options. Diurnal variance, the range of temperatures within a day, is generally also harmful. Crops are harmed by wide swings of temperature within the day. However, diurnal variance in the autumn is beneficial as many crops use diurnal variance as a signal for ripening.

Another issue raised in early criticisms of the original Ricardian paper concerns water. Although the original Mendelsohn et al. (1994) analysis captured sources of water from within a county, there was no data on water coming from outside the county. Farmers near rivers would have an advantage over farmers that were far from rivers. By omitting surface water supplies, the original Ricardian studies might have been biased. In order to test this potential source of bias, a subsequent analysis included available surface water by county (Mendelsohn and Dinar, 2003). Farms in counties with surface water resources that originate outside the county had higher values. However, including surface water withdrawals did not change the climate coefficients significantly, suggesting the bias associated with not having water data is small in the US.

Even if inclusion of surface water sources does not matter a great deal in the US, there is nonetheless a big difference between irrigated and rainfed cropland. One study of the US compared a regression with all observations with a regression that included only counties with rainfed agriculture (Schlenker et al., 2005). The study shows that the climate coefficients from the sample of just rainfed farms are significantly different from the climate coefficients in the regression that includes all farms. The results suggest that rainfed and irrigated agriculture do not react to climate in the same way; rainfed farms are more temperature sensitive than the irrigated farms. The authors suggest that future analyses should estimate rainfed and irrigated farms separately. Of course, this cannot be done with US data because net revenues for rainfed and irrigated farms are not reported separately. The authors further argue that Ricardian analyses of the entire US would be biased. However, the fact that rainfed and irrigated farms respond differently to climate does not imply that an analysis of all farms would be biased. It merely implies that an analysis of all farms would poorly predict the impact to just rainfed farms. Similarly, an analysis of rainfed farms would poorly predict the impact to all farms.

Another important analysis of the US agricultural sector reveals that the Ricardian results may not be stable over time (Deschenes and Greenstone, 2007). Using data from 1978, 1983, 1988, 1993, 1998 and 2003, the results suggest a general trend of increasing temperature sensitivity over time. The results suggest that there are missing variables from the cross-sectional Ricardian analyses that fail to predict how the regression coefficients might change over time. Unfortunately, the authors do not study the issue further to see if they can determine what these variables may be.

BRAZIL AND INDIA

The first Ricardian studies of climate change in developing countries focused on Brazil and India precisely because both of these countries kept good agricultural records (Dinar et al., 1998; Mendelsohn and Dinar, 1999; Mendelsohn et al., 2001; Kumar and Parikh, 2001; Sanghi and Mendelsohn, 2008). These studies use the Ricardian method to examine the net revenue (India) and land value (Brazil) in each district or municipio, respectively. Net revenue or land value is regressed on climate, soils and other control variables.

The Indian and Brazilian farm data provide a long-term panel. In order to get a long-term economic response from this data, district/municipio average farm values over a decade were averaged. This led to two periods of data for Brazil, 1970–75 and 1980–85, and two periods for India, 1966–75 and 1977–86. These long time periods allow testing whether the Ricardian responses are stable or changing.

Alternative functional forms were tried. The loglinear functional form appeared to fit the data most closely. The loglinear form has two advantages. First, it controls for the heteroscedasticity of the errors. Observed farm values and net revenues per hectare tend to be skewed to the right. Second, it postulates that climate change impacts are proportional to the value of the land rather than additive (such as in the linear case).

Because land values were not available in India, it was necessary to rely on annual net revenues as a measure of economic productivity. Net revenues are gross revenues minus costs. The rental rate for capital was calculated as a cost using the market value of each item and its expected length of service. One positive aspect of net revenue is that it does not include potential speculative aspects of land value. However, a limitation of annual net revenue is that it reflects the outcomes specific to only one year. Events, such as that year's weather, that are particular to that year are reflected in annual net revenue. This does not necessarily lead to bias but it does lead to a noisier measurement of long-term economic value compared to land value. Therefore, when using net revenue per year, the analyst has to be careful to select a representative weather year and not an abnormal weather year.

As discussed in Chapter 4, the reliance on net revenue in a developing country context raises two additional important points. First, many farmers in developing countries use their own family's labor. Since they do not pay themselves, it is not possible to get a reliable measure of this cost. Hired wage labor rates tend to be specific to the harvest season and do not represent wages during the entire year. We consequently omit own labor as a cost in the definition of net revenue. Second, many families consume a sizeable fraction of their output. We include this own consumption in our definition of net revenue and value it at market prices. Another issue with developing countries is the use of animals for farm operation. Animals were included, valued at their market price.

The independent variables included in the Indian regression include the same seasonal climate variables as used in the US. Other control variables include literacy, latitude, population density, high yield varieties, bulls/hectare, tractors/hectare, and six soil variables. The districts were weighted by the acreage of cropland to place more emphasis on where cropland is grown.

The results of the Indian Ricardian regression on net revenue are shown in Table 7.2 (Mendelsohn et al., 2001). The seasonal temperature and precipitation coefficients in both periods are significant in every season. Many of the quadratic terms are also significant, implying that the climate effects are non-linear. The control variables are also significant as expected. Literacy, bulls/hectare, and tractors/hectare all have positive expected impacts. The different soils variables affect net revenues. Larger farms lead to lower values. Because the cost of own labor is not included, net revenue inflates the value of small farms which rely heavily on household labor. One unexpected result is that net revenues are lower in areas with higher population density. It was expected that population density was a proxy for market access and so would be positive. Also surprisingly, the use of high yield varieties had no effect on net revenues.

The results of the Brazilian Ricardian regression are shown in Table 7.3 (Mendelsohn et al., 2001). The dependent variable in the Brazilian regression is farm land value per hectare. The regression includes four seasons but

Table 7.2   Indian Ricardian panel regression

Variable	1966–75	1977–86
Jan. temp.	9	–32
Jan. temp. sq.	–2.85	11.4*
Apr. temp.	–144**	–102*
Apr. temp. sq.	–16.4*	–8.8
Jul. temp.	–121*	–299**
Jul. temp. sq.	–23.3**	–6.0
Oct. temp.	–16	353**
Oct. temp. sq.	–21.6	12.9
Jan. prec.	–10.0**	18.5**
Jan. prec. sq.	–0.076	–0.227*
Apr. prec.	–2.30	–9.81**
Apr. prec. sq.	–0.055*	0.079*
Jul. prec.	–1.40**	–1.16**
Jul. prec. sq.	0003**	0.002**
Oct. prec.	6.94**	3.85**
Oct. prec. sq.	–0.030**	0.005
Control variables		
1966–76	5691**	
1977–86	5189**	
Population density	–28.1**	
Literacy	119**	
Latitude	49.0*	
High yield	–4.86	
Bulls/ha	106*	
Tractors/ha	2095	
Farm size	–146**	
Soil 1	303**	
Soil 2	193**	
Soil 3	–123*	
Soil 4	–27	
Soil 5	90*	
Soil 6	103*	

Notes:
$N = 5624$ and Adjusted $R^2 = 0.93$.
Dependent variable is log of farm net revenue per ha.
Coefficients significant at 5% level are marked with one asterisk (*) and at 1% level with two asterisks (**).
Climate coefficients allowed to vary over time.

Source: Mendelsohn et al. (2001, Table 4).

Table 7.3   Brazilian Ricardian panel regression

Variable	1970–75	1980–85
March temp.	99.2**	90.0**
March temp. sq.	9.47*	7.13*
June temp.	–131.0**	–123.0**
June temp. sq.	17.4**	12.5**
Sept. temp.	146.0**	172.0**
Sept. temp. sq.	–36.8**	–36.1**
Dec. temp.	–164.0**	–182.0**
Dec. temp. sq.	2.35	14.2**
March prec.	2.07**	2.44**
March prec. sq.	0.24	–4.37**
June prec.	2.95**	1.72**
June prec. sq.	–3.32	–0.22
Sept. prec.	–0.051	0.92
Sept. prec. sq.	–47.5**	–52.3**
Dec. prec.	–3.02**	–2.75**
Dec. prec. sq.	4.20*	0.02
Control variables		
Soils 1	–28	
Soils 2	140**	
Soils 3	–422**	
Soils 4	–72	
Soils 5	–479**	
Soils 6	408**	
Soils 7	143**	
Soils 8	–275**	
Soils 9	125**	
Soils 10	714**	
Soils 11	553**	
Soils 12	587**	
Latitude	–113**	
Population (000)	390**	
Population sq.	390**	
Year 70	–1506**	
Year 75	–285**	
Year 80	–15	

Notes:
N=14827 and Adjusted R^2 = 0.76.
Coefficients multiplied by 1000.
Coefficients significant at 5% level marked with one asterisk (*) and at 1% level with two asterisks (**).
Dependent variable is log of farmland value.
Climate coefficients are allowed to vary over time.

Source: Mendelsohn et al. (2001, Table 5).

they are characterized by the months of March (autumn), June (winter), September (spring), and December (summer). These months capture the center of the rainy season in Brazil. Additional control variables include latitude, population, population squared, and 12 soil variables. Dummies for each year are included to remove the effects of inflation and other year effects. The municipios were weighted by hectares of cropland.

Agriculture in Brazil is sensitive to both temperature and precipitation. The seasonal coefficients are significant in all seasons. The control variables also had significant roles. Soils explained some of the variation in land values across Brazil. The further a location is south of the equator, the higher are the land values. Larger populations in a municipio increased land values at a decreasing rate.

It is interesting to compare the climate coefficients in the early and later periods for both countries. In Brazil, most of the linear climate coefficients are stable except for the positive coefficient on winter precipitation which became smaller over time. Many of the squared terms, however, changed over time. The winter temperature squared term decreased and the summer temperature squared term increased. The fall precipitation squared term became more negative and the summer precipitation squared term became more positive.

There were many more changes in the Indian results over time. The linear summer temperature term became more negative and the linear autumn term more positive. The squared temperature term for winter went from negative to positive and the summer temperature squared term became less negative. The linear winter term for precipitation went from negative to positive whereas spring precipitation became more negative and autumn less positive. The quadratic precipitation term for winter became more negative, whereas the spring squared term went from negative to positive and the autumn term went from negative to zero.

The aggregate data studies of both India and Brazil (Mendelsohn and Dinar, 1999; Mendelsohn et al., 2001; Kumar and Parikh, 2001; Sanghi and Mendelsohn, 2008) used interpolated climate data from nearby weather stations. Although the weather stations take accurate recordings of weather over time (climate), they are not always located close to farms. Subsequent studies consequently examined the potential of satellites to provide both temperature and precipitation measures (Mendelsohn et al., 2007a). The advantage of satellites is that they can take direct measurements of the entire earth. One of the disadvantages of satellites is that they cannot measure everything of importance accurately. Direct comparisons revealed that at least the polar orbiting US Department of Defense Satellites provide more accurate measures of temperature than interpolated weather stations data (Mendelsohn et al., 2007a). In contrast, the available satellites cannot measure precipitation directly. The best proxy for precipitation is soil moisture but this did not work as well as interpolated precipitation. The results suggest that the best climate measurements would be a combination of temperature data from satellites and precipitation data from ground weather stations (Mendelsohn et al., 2007a).

In a subsequent analysis of the Indian and Brazilian data, Sanghi and Mendelsohn (2008) pool the data set across the entire time period. This does not provide a test of how coefficients change over time but it provides an accurate measure of the climate coefficients themselves. The dependent variable in the regression is net revenue per hectare. Figure 7.5 (see p. 126) presents a map showing how a warming of 2.0°C with a 7 percent precipitation increase would affect net revenues across India. The map provides valuable observations about the pattern of impacts across India. The impacts in the East and North are beneficial but the impacts in the West are very harmful. For example, Gujarat has districts that will lose between 600 to 2800 US dollars per hectare whereas eastern states, such as Orissa, Andhra Pradesh and Tamil Nadu, may face benefits ranging from 0 to 1500 US dollars per hectare. Another important observation revealed in the map is that many inland states face a significant variation in the impact across the various districts. Both the inter-state and intra-state variations in climate impact are a major policy concern and will be discussed in Chapter 12.

Figure 7.6 (see p. 127) is a map for Brazil using the same climate scenario of 2.0°C warming with a 7 percent precipitation increase. The Brazilian analysis is also using pooled data to get a single best estimate of climate effects. The dependent variable in the Brazilian regression, however, is the log of land value. The map shows the percentage change in land value with this climate scenario. The results reveal that impacts in Brazil also vary across space. The southeastern part of Brazil may even benefit from warming. However, the north and west of Brazil are much more likely to be harmed. The map reveals that roughly 90 percent of the agricultural areas of Brazil faces a loss in land value of between between 8 percent and 100 percent compared with the present climate. Northeastern states such as Ceara, Maranhao and Tocantins face the most significant negative impact, ranging between 30 percent to 100 percent reduction in land value. Southeast states such as Rio Grande do Sul and Santa Catarina benefit greatly, possibly by as much as 40–50 percent.

SRI LANKA

In contrast to the studies reviewed to date, we now review another example of aggregate data analysis for a small country, Sri Lanka (Seo et al., 2005).

Table 7.4   Sri Lanka Ricardian regression

Variables	Coefficients
May temp.	–4105*
January rain	103*
September rain	93*
November rain	–141*
Irrigated acreage	–0.15
Altitude	27.3*
Adjusted R^2	0.59

Notes:
N = 25.
The dependent variable is net revenue per hectare.
Coefficients significant at 5% level are marked with one asterisk (*).

Source: Seo et al. (2005).

Net revenue per hectare is regressed against climate and control variables across 25 districts in the country. The limited number of observations limits the complexity of the model that can be estimated. In the case of Sri Lanka, there is not a great deal of temperature variation across the country, although there are significant differences in rainfall patterns. The study consequently focuses on a relatively simple linear model of May temperature with a few seasonal precipitation variables for January, September and November. Other control variables include irrigated acreage and altitude.

The coefficient on May temperature reveals that higher temperatures would lead to a reduction in net revenues in Sri Lanka (Table 7.4). Higher annual rainfall levels would increase net revenue. However, seasonal changes would not have the same consistent effect. Increased rain in January and September is beneficial, but more rain in November is harmful.

Having more irrigated land surprisingly reduces net revenue in Sri Lanka. In other countries, irrigated land is more valuable than rainfed land. However, rainfed farms in Sri Lanka have ample rain and grow high-value crops such as tea. Higher altitude farms earn less revenue. Higher altitudes are generally stressful on crops and increase costs in every country.

COMPARISON OF MARGINAL IMPACTS

In order to get a better sense of the climate results in the US, Brazil, India and Sri Lanka (Tables 7.1 to 7.4), we examine the marginal impact of

*Table 7.5   Marginal impact of temperature and precipitation across
            countries*

Country	Temperature °C	Precipitation mm/mo
United States		
1978 cropland	−19.2	4.2
1982 cropland	−12.0	3.9
1978 crop revenue	−9.5	7.5
1982 crop revenue	−2.9	4.6
India		
1966–75	−242.7	−60.3
1977–86	−36.4	51.7
Brazil		
1970–75	−41.3	1.6
1980–85	−19.4	1.1
Sri Lanka		
1995	−49.9	0.7

Notes:
Values measured in 2008 USD/ha/yr.
Values updated with US GDP deflator.
Annual values calculated assuming a real interest rate of 5%.
Values calculated at mean climate of each country using climate coefficients from earlier
tables in the chapter.

Sources: Mendelsohn et al. (1994, 2001); Seo et al. (2005).

temperature and precipitation in Table 7.5. The data from each of the
aggregate studies was demeaned. The linear coefficients of the climate variables consequently reflect the marginal impacts at the mean climate. All
marginal values are expressed in USD/ha/yr per degree Centigrade of temperature or mm/mo of rainfall. An interest rate of 5 percent has been used
to convert land values to net revenue per year to make them comparable.
Marginal values have been updated to 2007 using the US GDP deflator.

The results in Table 7.5 suggest that increased temperature is harmful
to agriculture in every country. The impacts on the US are the smallest
and the impacts on India are the largest. These cross-country results are
consistent with a general pattern that net revenue (or land value) has a hill-
shaped relationship with temperature, reaching a maximum in the temperate zone and declining thereafter. The warmest countries consequently
have the largest negative temperature effects. More precipitation is generally beneficial, although there is one exception in the early period of India.
India is much more sensitive to precipitation than all the other countries,
partly because it is so dry in the non-monsoonal periods.

For the US, India and Brazil, Table 7.5 reveals that the climate coefficients are not stable over time (a result also found by Deschenes and Greenstone, 2007). Temperature sensitivity declines in all three countries over the tested period (Deschenes and Greenstone, 2007 find climate sensitivity increases). The positive precipitation coefficients in Brazil and the US shrink over time. In India, precipitation shifts from being harmful to being beneficial. It is not clear how much to read into the fact that the Ricardian coefficients are not stable over time; it is likely that there are missing factors that change over time that are causing this shift. Future research should determine what could be missing from the cross-sectional models to make them more reliable over time.

CONCLUSION

Aggregate data has been a fertile ground for Ricardian models. In the countries where such data exists, there has often been sufficient variation in climate across districts, counties and municipios to conduct cross-sectional analyses. These analyses reveal that farm income is indeed sensitive to climate. Higher temperatures tend to be only slightly harmful in temperate locations but increasingly harmful in warmer settings. More precipitation has a positive but diminishing beneficial effect.

There have been a number of technical improvements to the application of the Ricardian approach to aggregate data. Satellite temperature data has replaced interpolated ground station data in places with sparse measurements. Alternative functional forms have been explored. Missing control variables, especially water, have been tested. Ricardian analyses of all farms have been contrasted with analyses of just rainfed farms. Finally, several studies have examined whether the Ricardian functions are stable over time.

These studies have answered many questions, but in turn, have raised many more. Surface water, for example, is clearly valuable in dry locations. Yet it is not clear whether surface water supplies must be included in the analyses to get unbiased estimates of climate coefficients. The loglinear functional form has promising attributes. Climate variables should be introduced in a non-linear fashion. The initial characterization of climate used a single month to represent each season. However, not every study used the same month. Rainfed farms are more sensitive to temperature than all farms together. This suggests that irrigated farms are less sensitive than rainfed farms to temperature, but this cannot be directly tested with available aggregate data. The Ricardian climate coefficients change over time. Does this reflect a weakness in the technique or are there missing variables that would explain this shift if included? What variables are changing over time that would explain the observed shifts? Are there changes in technology, prices, or other variables that are not yet captured? All these questions suggest that future additional work is important and needed.

REFERENCES

Deschenes, O. and M. Greenstone (2007), 'The economic impacts of climate change: evidence from agricultural output and random fluctuations in weather', *American Economic Review*, **97**, 354–85.

Dinar, A., R. Mendelsohn, R.E. Evenson, J. Parikh, A. Sanghi, K. Kumar, J. McKinsey and S. Lonergan (1998), 'Measuring the impact of climate change on Indian agriculture', World Bank Technical Paper 402, Washington, DC, USA: World Bank.

Hadley Centre (2008), 'Climate projections', United Kingdom Meteorological Office, available at www.metoffice.gov.uk/research/hadleycentre/models/modeldata.html.

Kumar, K. and J. Parikh (2001), 'Indian agriculture and climate sensitivity', *Global Environmental Change*, **11**, 147–54.

Massetti, E. (2009), 'Essays on the economics of mitigation and adaptation to climate change', PhD thesis, Università Cattolica del Sacro Cuore, Milan, Italy.

Mendelsohn, R. and A. Dinar (1999), 'Climate change impacts on developing country agriculture', *World Bank Research Observer*, **14**, 277–93.

Mendelsohn, R. and A. Dinar (2003), 'Climate, water, and agriculture', *Land Economics*, **79**, 328–41.

Mendelsohn, R., A. Dinar and A. Sanghi (2001), 'The effect of development on the climate sensitivity of agriculture', *Environment and Development Economics*, **6**, 85–101.

Mendelsohn, R., W.D. Nordhaus and D. Shaw (1994), 'Measuring the impact of global warming on agriculture', *American Economic Review*, **84**, 753–71.

Mendelsohn, R., W.D. Nordhaus and D. Shaw (1996), 'Climate impacts on aggregate farm value: accounting for adaptation', *Agricultural and Forest Meteorology*, **80**, 55–66.

Mendelsohn, R., W.D. Nordhaus and D. Shaw (1999), 'The impact of climate variation on US agriculture', in R. Mendelsohn and J. Neumann (eds), *The Impact of Climate Change on the United States Economy*, Cambridge, UK: Cambridge University Press.

Mendelsohn, R., A. Basist, A. Dinar, F. Kogan, P. Kurukulasuriya and C. Williams (2007a), 'Climate analysis with satellites versus weather station data', *Climatic Change*, **81**, 71–84.

Mendelsohn, R., A. Basist, A. Dinar and P. Kurukulasuriya (2007b), 'What explains agricultural performance: climate normals or climate variance?', *Climatic Change*, **81**, 85–99.

Quiggin, J. and J. Horowitz (1999), 'The impact of global warming on agriculture: a Ricardian analysis: comment', *American Economic Review*, **89**, 1044–45.

Sanghi, A. and R. Mendelsohn (2008), 'The impacts of global warming on farmers in Brazil and India', *Global Environmental Change*, **18**, 655–65.

Schlenker, Wolfram, Michael Hanemann and Anthony Fischer (2005), 'Will US agriculture really benefit from global warming? Accounting for irrigation in the hedonic approach', *American Economic Review*, **95**, 395–406.

Seo, S.N., R. Mendelsohn and M. Munasinghe (2005), 'Climate change and agriculture in Sri Lanka: a Ricardian valuation', *Environment and Development Economics*, **10**, 581–96.

US Census Bureau (1984), *1982 Census of Agriculture*, Washington, DC, USA: US Government Printing Office.

US Census Bureau (1988), *County and City Data Book*, Washington, DC, USA: US Government Printing Office.

Washington, W., J. Weatherly, G. Meehl, A. Semtar, B. Bettge, A. Craig, W. Strand, J. Arblaster, V. Wayland, R. James and Y. Zhang (2000), 'Parallel Climate Model (PCM): control and transient scenarios', *Climate Dynamics*, **16**, 755–74.

8.   Ricardian models of individual farms

In order to extend the Ricardian analyses to countries that do not have existing agricultural census data, it is necessary to collect information about farms across climate zones. Such surveys are most easily done at the individual farm level. The collected data provides more detailed information about farm activities and the resulting income from each activity. For example, individual surveys permit separate analyses of livestock, rainfed crops and irrigated crops.

By definition, farm income or net revenue per hectare is gross revenue minus costs. The survey instrument needs to collect data about individual products sold and their prices in order to calculate farm-level gross revenue (the sum of revenue from all sources). Because the purpose of the Ricardian analysis is to measure the overall productivity of the land, gross revenue also includes the value of products consumed by the household. These products are valued at the market price. Because both livestock and crop products can be measured separately, it is possible to measure gross crop revenue and gross livestock revenue. Subtracting the costs of each activity yields crop net revenue and livestock net revenue. Irrigated plots lead to irrigated net crop revenue and rainfed plots lead to rainfed net revenue.

In order to measure costs, the survey instrument collected data on the costs of farm production (see Dinar et al., 2008). For crops, this includes information on variable costs such as seeds and fertilizer as well as fixed costs such as machinery, equipment and animal power. For animals, this includes the cost of the stock of animals owned as well as purchased feed. However, in many farms in these samples, the animals relied strictly on natural vegetation for sustenance. Fixed (capital) costs are converted to annual costs using equation (4.10) in Chapter 4. The equation determines the annual rent required to use the equipment for a year. An annual real interest rate of 5 percent was assumed and the depreciation rate varied with the expected lifetime of the equipment. Some costs were difficult to calculate. Household labor was possible to measure in hours but difficult to value, since households do not pay themselves. Hired wages were found to be non-representative because laborers were hired largely during the harvest season when wages are high. The omission of household labor costs led to a difference in net revenues between small and large farms. Household labor is a larger fraction of small farm costs so that small farms appear to have slightly higher returns per hectare than large farms in the Ricardian studies. Another cost that was not taken into account is irrigation. The cost of building irrigation systems is difficult for farmers to assess, given that many of the systems are built over time and some are built by the government. Irrigation is another cost left out of the net revenue calculation. The omission of irrigation costs inflates the net revenue of irrigated land relative to rainfed land. We report the design and results of two regional studies in Africa and South America. The African study involved surveys of 11,000 farms across 11 countries. The South American survey collected data from 2500 farms across seven countries.

AFRICA

The first study that collected individual farm data to estimate a Ricardian function was an analysis of climate effects on farmland in Africa. The analysis examined both crop and livestock net revenue. A survey was designed and then pretested to determine whether the instrument would be effective across Africa (Dinar et al., 2008). Teams from Burkina Faso, Cameroon, Egypt, Ethiopia, Ghana, Kenya, Niger, Senegal, South Africa, Zambia and Zimbabwe then administered the surveys to farmers in their own country. Care was taken in the sample design to select farms from across all the climate zones within a country that support agriculture. Consequently, there is significant variation in climate across farms within countries as well as between countries. The economic data was matched with temperature data from satellites and precipitation data that was interpolated from weather stations. Soils from FAO soils maps were matched with each farm. Finally, natural water flows were calculated using a hydrological model.

Crop net revenue was regressed against climate, soils and other control variables (Kurukulasuriya et al., 2006; Kurukulasuriya and Mendelsohn, 2008) using a linear functional form with quadratic climate variables. Separate regressions are estimated for all farms and for the subsample of irrigated farms and the subsample of rainfed farms (see Table 8.1). The comparison of the results from the two subsamples allows a direct test of whether rainfed and irrigated plots have similar climate sensitivity. Regressions included regional variables to capture major differences between countries such as language and religion. The R-squared statistics for the three regressions varies from 0.2 to 0.4. These values are much

Table 8.1   African Ricardian regression

Variable	All farms	Rainfed	Irrigated
Winter temp.	–173.6**	–106.7	–93.5
Winter temp. sq.	6.1**	3.9*	4.9
Spring temp.	115.1	–82.8	58.7
Spring temp. sq.	–5.0**	–0.3	–4.1
Summer temp.	173.9**	198.6**	827.5**
Summer temp. sq.	–1.9	–3.2*	–13.1*
Autumn temp.	–98.1	–92.4	–824.2*
Autumn temp. sq.	1.1	1.5	15.3*
Winter prec.	–2.9*	–1.9	5.8
Winter prec. sq.	0.0**	0.00	0.00
Spring prec.	3.5*	3.6**	–10.6
Spring prec. sq.	–0.001	–0.011*	0.091*
Summer prec.	3.4**	1.9*	21.4**
Summer prec. sq.	–0.012**	–0.005	–0.086**
Autumn prec.	–0.5	–0.6	–14.7**
Autumn prec. sq.	0.0055*	0.0053*	0.0586**
Mean flow	9.4**	–5.4	8.8**
Farm area	–0.1**	–0.3**	–0.0**
Farm area sq.	0.0*	0.0**	0.0*
Elevation	0.035	–0.0009	0.229
Log (household size)	22.9	10.1	62.4
Irrigate(1/0)	237.5**		
Electricity (1/0)	66.6**	47.7**	233.2*
Eutric Gleysols: *coarse, undulating*	–631**	–287**	–540
Lithosols and Luvisols: *hilly to steep*	–387**	–156**	–1147**
Orthic Luvisols: *medium, hilly*	–2181**	–1959**	
Chromic Vertisols: *fine, undulating*	–1180**	–1006**	–1719**
Chromic Luvisols: *medium to fine, undulating*	–295**	–241**	
Cambic Arenosols	1633**	1726**	
Luvic Arenosols	–482**	–188**	
Chromic Luvisols: *medium, steep*	–2153		–6157**
Dystric Nitosols	214		7051**
Gleyic Luvisols	–199**	–154**	
Rhodic Ferralsols: *fine, hilly to steep*	1428**		3212
Calcic Yermosols: *coarse to medium, undulating to hilly*	1071**	148	
West Africa dummy	136**	208**	–285
North Africa dummy	457**		675*
East Africa dummy	–186**	–154**	–361
Heavy machinery dummy	51.8**	55.5**	–60.8
Animal power dummy	10.4	49.3**	–185.5**

Table 8.1   (continued)

Variable	All Farms	Rainfed	Irrigated
Constant	−388	1081	−549
N	8459	7238	1221
R²	0.40	0.20	0.30
F-test	63.6	32.4	46.3

Notes: Values significant at 5% level are marked with one asterisk (*) and values significant at 1% level are marked with two asterisks (**).

Source: Kurukulasuriya and Mendelsohn (2008).

lower than the results from aggregate data studies, but that is because there is a lot of variation at the individual farm level that is averaged away in the aggregate data. For all three regressions, the climate coefficients are highly significant.

Both temperature and precipitation have a significant effect on crop net revenue. The effects vary by season. The quadratic terms are significant. The climate coefficients in the irrigated and rainfed regressions are significant and they are different from one another. Mean water flow is also significant and positive in the regressions of all farms and irrigated farms. However, mean flow is not significant in the regression of rainfed farms, as expected.

The control variables are also clearly significant. Small farms have larger net revenue per hectare than larger farms. This most likely reflects the omission of household labor from net revenue. However, it may also be possible that small farms are more efficient per hectare. Farms with electricity have higher net revenue. This may reflect that they are more modern or that they are closer to urban areas and markets. At least some soil types have significant effects on farm net revenues. Rainfed North African farms have higher net revenues, but rainfed East African farms have lower net revenues. North African (Egyptian) irrigated land earns more net revenue than irrigated land in the rest of Africa. The dummy variables for heavy machinery and animal power increase the value of rainfed farms. The dummy variable for animal power decreases the net revenue from irrigated land. On irrigated land, the presence of animal power may reflect low technology, whereas on rainfed land, the lowest technology has neither animal power nor machinery.

In order to interpret the climate coefficients, we calculate the marginal effects and elasticities of the climate variables in Table 8.2. The results reveal that warming would reduce the net revenues of rainfed farms in

Table 8.2   Marginal African climate impacts

	All farms	Irrigated farms	Rainfed farms
Annual temperature	–28.5**	35.04	–26.7**
	(–1.4**)	(0.6)	(–1.9**)
Annual precipitation	3.28**	3.82	2.70**
	(0.44**)	(0.13)	(0.63**)

Notes:
Values calculated for Table 8.1 at mean climates.
Elasticities are in parenthesis.
Values significant at 5% level are marked with one asterisk (*) and values significant at 1%
level are marked with two asterisks (**).

Source: Kurukulasuriya and Mendelsohn (2008).

Africa. In contrast, warming would increase the net revenues from irrigated lands, although this effect is not significant. Increased precipitation would increase net revenues in irrigated and rainfed farms but the effect is significant only for rainfed farms. The elasticities suggest that the net revenues of rainfed farms are more sensitive to changes in temperature and rainfall than the net revenues from irrigated farms. This is expected given the moderating effect of having alternative water supplies on irrigated land.

In order to see how the marginal effects are distributed across the landscape, we present maps of the temperature and precipitation effects. Figure 8.1 (see p. 128) measures the marginal impact of a 1°C warming across Africa. Warming would have immediate strong negative impacts on west Africa and in many places in east Africa. However, northern Africa, including the Sahara, would benefit. Figure 8.2 (see p. 129) presents the marginal impact of a 1mm/month increase in precipitation. The desert regions would benefit the most. Most of the rest of Africa would also benefit except for the narrow range in central Africa that is already very wet.

In addition to measuring the marginal effect of a change in climate, it is also useful to measure the impact of possible future climate scenarios. Two predicted scenarios are presented: the change in climate by the Parallel Climate Model (PCM) (Washington et al., 2000) and by the Canadian Climate Centre (CCC) (Boer et al., 2000). The PCM predicts a relatively mild and wet climate scenario for Africa whereas the CCC model predicts a relatively hot and dry climate scenario. The two predicted scenarios reflect the wide range of scenarios considered plausible by the Intergovernmental Panel on Climate Change (IPCC, 2007). These

Table 8.3   Impacts of climate scenarios on African cropland

Impacts	PCM 2100	CCC 2100
Rainfed		
ΔNet revenue (USD/ha)	+161	–137
	(50.8%)	(–43%)
ΔTotal net revenue (billions USD)	+50.8	–43.5
Irrigated		
ΔNet revenue (USD/ha)	+352	+341
	(30%)	(29%)
ΔTotal net revenue (billions USD)	+4.6	+4.4
All Farms		
ΔNet revenue (USD/ha)	+213	–144
	(48%)	(–33%)
ΔTotal net revenue (billions USD)	+69.2	–46.8

Source: Kurukulasuriya and Mendelsohn (2008).

simplistic scenarios examine only the effect of a change in climate. Other factors that are likely to change over the century are not included, such as changes in prices, land use, technology and population. The scenarios are intended only to give a sense of how climate change will affect agriculture, and they should not be interpreted as forecasts of the future.

The results are shown in Table 8.3. The mild PCM scenario leads to increased net revenues in both irrigated and rainfed farms. African crops will do better if temperatures do not warm very much but precipitation increases. In contrast, in the CCC scenario, net revenues on rainfed farms will fall. Interestingly, the model suggests that irrigated farms will not be negatively affected unless water flows are reduced. The hotter, dryer CCC scenario will reduce African crop net revenues overall.

In order to understand how these climate effects are distributed across Africa, we present several maps of the results. Figure 8.3 (see p. 130) presents the map of the impact of the PCM climate scenario on African crop net revenues. Most of Africa is predicted to benefit in this scenario with only a few exceptions: a narrow sliver of land along the northwestern coast of Africa, an extension of land from west Africa through central Africa, and a large fraction of south Africa. Figure 8.4 (see p. 131) presents the impacts of the CCC climate scenario on African crop net revenues. Farmers from many regions of Africa will lose net revenue in this scenario. The only beneficiaries in this scenario are in west Africa and central Africa.

In addition to crops, it was also possible to examine the effect of climate on livestock (Seo and Mendelsohn, 2008a). Because livestock operations in small household versus large commercial farms are quite different, climate coefficients for small and large livestock farms were calculated separately. A small livestock farm was defined as owning less than 500 USD in livestock. However, in practice, small farms averaged about 250 USD of livestock owned, whereas large farms averaged almost 9000 USD of livestock owned. So the two types of livestock farms are quite distinct.

Unfortunately, it was not possible to measure the amount of land that each farm uses for livestock operations in most of Africa. African livestock owners rely largely on common grazing land for their animals. There is no specific amount of land per farmer. The standard Ricardian model, estimating impacts on a per hectare basis, could not be applied. In order to study climate effects, two dependent variables were studied: the value of animals owned per farmer and the net revenue per animal owned. Animals were combined into a single amount of stock by multiplying the number of each species owned by the price of each animal. Net revenue was calculated by multiplying the products from all animals by the price of each product.

Two African livestock regressions are shown in Table 8.4. The number of animals owned regression is identified using the percentage grassland in that district. The percentage grassland is a measure of the natural ecosystem, not man-made pastureland. Given that most African livestock feed directly from the landscape, this variable determines the number of animals the land can support. The climate coefficients in Table 8.4 reveal that the value of animals owned is sensitive to temperature and precipitation. Further, the climate coefficients of small and large farms are different. The control variables are also significant. Larger households and Muslim farmers have fewer animals. Farmers with electricity and farms located in areas with grassland and higher population densities have more animals.

The regression of the net revenue per animal owned is also significant. Climate variables have a significant effect on livestock net revenue. Both temperature and precipitation play a role. Small and large farms have different climate coefficients. Electricity and higher population densities lead to higher net revenue.

The marginal impact of climate on African livestock ownership and net revenue are calculated in Table 8.5. Small farms have relatively larger herds in warmer places. Small household farms are likely to shift from crops to livestock as temperatures warm (Seo and Mendelsohn, 2008a). They consequently have more options than commercial livestock farms that are livestock-only operations. Further, large livestock farms rely heavily on beef cattle that must be in temperate settings (Seo and Mendelsohn, 2006, 2008b). Because they are more specialized, large livestock farms

Table 8.4 African livestock regression

Variable	Value of livestock owned		Net revenue per livestock value in USD 1000	
	Coef.	t-stat.	Coef.	t-stat.
Intercept	12460	1.86	1424	6.72
Temp. × small[1]	−1049	−1.71	−49.9	−2.53
Temp. sq. × small[1]	28.2	2.10	0.55	1.28
Prec. × small[1]	−103	−2.98	−13.41	−12.05
Prec. sq. × small[1]	0.47	2.60	0.07	13.17
Temp. × large[1]	1351	7.15	14.90	2.43
Temp. sq. × large[1]	−42.8	−7.21	−0.50	−2.59
Prec. × large[1]	−7.62	−0.20	−2.67	−2.19
Prec. sq. × large[1]	−0.32	−1.47	0.01	1.07
Log household size	−2240	−4.55	10.57	0.66
Electricity dummy	4960	7.13	219.5	9.72
Population density	126.6	2.77	11.55	7.96
Population density sq.	−2.13	−4.21	−0.12	−7.79
% Muslim	−4508	−3.02	−31.75	−0.75
% Grassland	22952	10.58		
			Adjusted R^2 = 0.20, N = 4763	

Note: [1]Coefficients of the interaction between temperature and size describe the climate sensitivity of each farm size.

Source: Seo and Mendelsohn (2008a).

have fewer alternatives than small livestock farms as temperatures warm. Both small and large livestock farms have larger herds in dryer places (Seo and Mendelsohn, 2008a). Small farms are more sensitive to precipitation than large farms because they shift to crops in wetter locations (Seo and Mendelsohn, 2008a). The results reveal that small and large farms have quite different climate sensitivities. Because they have more flexibility (substitution alternatives), small livestock farms are less vulnerable to climate change than large livestock farms.

Table 8.6 displays the impacts of the 2100 PCM and CCC climate scenarios on the net revenues of African livestock. The change in net revenues reflects the combined consequence of changes in stock and net revenue per animal. The table shows the effect of each scenario on both small and large livestock farms. Net revenue increases in both climate scenarios for small livestock farms but the effect is much greater in the PCM scenario. Net revenues fall for large livestock farms, and the effects are similar in both climate scenarios.

Table 8.5   Marginal effect of climate on African livestock

Types	Current livestock income ($/farm)	Marginal temperature impact ($/°C)	Marginal precipitation impact ($/mm)	Temperature elasticity	Precipitation elasticity
Value of livestock owned					
Small	259	256.8*	–41.0*	22.9*	–10.6*
Large	7795	–357.9*	–93.0	–1.04*	–0.73
Net revenue per livestock value					
Small	0.371	–0.024*	–0.004*	–1.51*	–0.63*
Large	0.394	–0.033*	–0.006*	–1.87*	–0.94*

Note: Values significant at 5% level are marked with one asterisk (*).

Source: Seo and Mendelsohn (2008a).

Table 8.6   Impact of climate scenarios on African livestock

	PCM scenario	CCC scenario
Small farm	+2034	+338
	(+323%)	(+54%)
Large farm	–831	–720
	(–26%)	(–23%)

Notes:
Effects are measured in USD/farm.
Percentage changes in parentheses.
All effects are significant at 1% level.

Source: Seo and Mendelsohn (2008a).

The results of the livestock scenarios can also be presented in maps. Figure 8.5 (see p. 132) presents the impacts of the PCM scenario for 2100 on African livestock net revenue of small livestock owners. The map predicts gains for small livestock farmers throughout Africa. Benefits are especially high in areas that are currently wet. Figure 8.6 (see p. 133) examines the same climate scenario for large livestock farms. Large farms in only the southern tier of Africa are expected to do well in the PCM scenario. Large livestock farms will be hurt in the rest of Africa and especially in the Congo and Zambia.

Figure 8.7 (see p. 134) presents the results for the CCC scenario for small livestock owners. Small owners in all of Africa benefit, especially in sub-Saharan Africa. The beneficial results in this map help mitigate some of the crop losses expected in this scenario for small farmers. There are especially large gains expected in Angola, Zambia and the Congo. Figure 8.8 (see p. 135) presents the corresponding results for large livestock farmers. None of the large livestock owners do well in the CCC climate scenario. The losses are especially large in Angola, Zambia and the Congo. It is quite clear that small and large livestock owners have very different sensitivities to climate. As discussed earlier, this is likely to be due to the specialization of large livestock owner in high-value but heat-intolerant breeds of cattle.

SOUTH AMERICA

The study of South America involved seven countries: Argentina, Brazil, Chile, Columbia, Ecuador, Uruguay and Venezuela (Mendelsohn et al., 2007). A survey was designed for South America using the African survey as a model, but because the detailed description of households proved to be ineffective, it was streamlined in the South America survey. Additional crops and livestock relevant to the region were added. As with the African study, teams conducted the surveys of farmers across a broad array of climates within each country in order to observe as much climate varia-tion as possible. A total of about 2500 surveys were administered. Of this total, 2283 of these farms raised crops. A total of 1753 of the farmers grew rainfed crops and another 530 farms grew irrigated crops. Of the 2500 total farms, 1278 raised livestock. Many of the small farms raised both crops and livestock.

In the Ricardian model, land value is regressed on climate, soils and other control variables (Seo and Mendelsohn, 2008c). The climate vari-ables include just winter and summer (not all four seasons). Climate inter-action terms between temperature and precipitation are also included. The control variables in the South American study include 10 soil vari-ables, altitude, electricity, a computer, age and sex of head of family, and country dummies. Individual country dummies are added to control for missing variables that vary from country to country. Three regressions are estimated. One is estimated across the entire sample of crop farms, one across only rainfed farms, and one across only irrigated farms.

In order to test whether small and large farms have similar climate sensitivity, a set of interaction terms is included. A dummy variable for large (commercial) farms is interacted with each climate variable. The coefficient on this interaction term tests whether large farms have different climate coefficients than small farms.

The results of the South America regressions are shown in Table 8.7. Six of the ten climate coefficients for the sample of all farms are significant. However, only precipitation coefficients are significant for the two subsamples. The precipitation coefficients for rainfed and irrigated farms are not similar. Only two of the ten climate coefficients for small and large farms are significantly different. For all farms, large farms have a more negative response to warmer winter temperatures and wetter summer precipitation than small farms. For rainfed farms, larger farms do better with warmer summer temperatures but worse with warmer winter temperatures compared to small farms. With irrigated farms, large farms have a positive coefficient on squared winter temperature and a negative coefficient on summer squared precipitation compared to small farms. The results suggest that there may be climate sensitivity differences between small and large farms, but they are subtle.

The regressions also reveal that soils matter. Every significant soil coefficient except the coefficient for Yermasols increases farm value for all farms. Luvisols and Vertisols are particularly valuable. Rainfed farms react particularly well to Phaezoms as well as Ferrasols and Vertisols. Although not significant, irrigated farms reacted well to Phaezoms and to Arenosols but not to Cambisols or Ferrasols. The results suggest that the value of soils depends a great deal on whether or not land is going to be irrigated.

Higher altitude is beneficial to rainfed crops but harmful to irrigated crops. It is very likely that the cost of water is lower in lower altitudes closer to rivers. Electricity is beneficial to farms, especially irrigated farms. This may reflect the need to pump water for irrigation. Computers are linked to higher net revenue. This may reflect their actual use in production, it may be an endogenous indicator of income, or possibly the computers reflect access to urban markets. Clay soils are deleterious to rainfed farms. Older household heads earn less money irrigating. This may reflect problems running an irrigated farm as one gets older or it may reflect the difference between old-fashioned and modern irrigation methods. Whether the head of the household is a female does not matter. Farmers in Argentina, Chile and Uruguay earn significantly less per hectare than Brazilian farmers but farmers in Columbia earn more. Education, though not shown, was found to be insignificant.

The marginal impact of temperature and precipitation is calculated in Table 8.8 for small and large farms in Latin America (South America plus Mexico, Central America and the Caribbean). The marginal impact of temperature is negative across the board. Small and large farms, rainfed and irrigated farms all are harmed by increasing temperature. The marginal effect of higher temperature is not significantly different for small and large farms. The marginal temperature effect is also not significantly different

Table 8.7   Ricardian regressions for all farms, rainfed farms and irrigated farms in Latin America

Variables	All farms	Rainfed farms	Irrigated farms
	Coef.	Coef.	Coef.
Intercept	2879*	1114	6707*
Temperature summer	285.1*	132.1	211.0
Temp. summer sq.	–10.31*	–5.30	–6.87
Temperature winter	–153.4*	41.72	–225.4
Temp. winter sq.	0.26	–3.98	–2.25
Precipitation summer	–4.12	6.74	–26.60*
Prec. summer sq.	–0.01	–0.02*	0.02
Precipitation winter	–9.38*	–7.34	–24.95*
Prec. winter sq.	0.01	0.03*	0.11*
Temp. summer × large	–43.3	216.6*	84.6
Temp. summer sq. × large	2.14	–4.79	–5.71
Temp. winter × large	–101.3*	–267.7*	–771.1
Temp. winter sq. × large	1.3	4.2	29.2*
Prec. summer × large	7.28	2.73	28.70
Prec. summer sq. × large	–0.02*	–0.01	–0.14*
Prec. winter × large	–4.8	–5.88	11.27
Prec. winter sq. × large	0	–0.03	–0.11
Temp. × prec. summer	0.37*	0.09	1.01
Temp. × prec. winter	0.63*	0.40*	0.40
Temp. × prec. summer × large	0	0.07	0.65
Temp. × prec. winter × large	0.39	0.55	1.68
Soil Cambisols	–3.03	–0.24	–14.33
Soil Ferrasols	12.77*	14.48*	–13.88
Soil Phaeozems	11.74*	22.89*	6.21
Soil Luvisols	27.07*	–1.61	20.82
Soil Arenosols	5.55*	2.70	16.26*
Soil Regosols	10.66*	8.82*	12.18
Soil Vertisols	16.01*	14.35*	9.09
Soil Yermosols	–2.5	0.62	–3.80
Altitude	0.18	0.49*	–0.75*
Electricity dummy	421.4*	214.2	1085.7*
Computer dummy	363.7*	439.4*	73.5
Texture (mixed)	–64.3	–172.6	862.7
Texture (clay)	–361.5*	–474.3*	416.0
Age of the head	–6.1	–0.1	–27.0*
Female dummy	–46.2	10.0	–103.2
Argentina	–1423*	–1369*	–2199*
Chile	–2356*	–1602*	–3822*
Columbia	804*	764*	2188*

Table 8.7   (continued)

Variables	All farms	Rainfed farms	Irrigated farms
	Coef.	Coef.	Coef.
Ecuador	−237	−460	92
Uruguay	−2534*	−2050*	
Venezuela	−40	233	−726.47
N	2283	1753	530
F-test	16.4	14.91	5.88
Adjusted R²	0.22	0.28	0.28

Notes: One asterisk denotes significance at 5% level.
The dependent variable is land value in USD per hectare of land in 2003–04.

Source: Seo and Mendelsohn (2008c, Table 1).

Table 8.8   *Marginal effects and elasticities for all farms, rainfed farms and irrigated farms in Latin America*

	Small farms Temperature	Large farms Temperature	Small farms Precipitation	Large farms Precipitation
All farms	−155.4*	−156.6*	1.36	4.49
	(−1.21*)	(−1.89*)	(0.07)	(0.35)
Rainfed farms	−101.3*	−170.4*	5.48*	3.46
	(−0.95*)	(−2.05*)	(0.35*)	(0.28)
Irrigated farms	−197.9*	−116.6	−12.54	25.33
	(−1.03*)	(−1.47)	(−0.33)	(1.68)

Notes:
Calculated from the coefficients in Table 8.7 at the mean of the corresponding samples.
One asterisk denotes significance at 5% level.
Marginals are changes in land value (USD) from temperature increase of 1°C or rainfall increase by 1 mm at the mean of the samples.
Elasticities (in parentheses) are corresponding percentage changes in land values.

Source: Seo and Mendelsohn (2008c).

for rainfed versus irrigated farms. The marginal precipitation effects are generally positive although not significant. The only marginal effect that is negative is for small irrigated farms and the only marginal effect that is significant is for small rainfed farms. Again, the difference between small and large farms is not significant, nor is the difference between rainfed and irrigated farms. The marginal estimates suggest that all the tested types of farms will respond to climate in a similar fashion.

The marginal impact of temperature on rainfed farms is presented in Figure 8.9 (see p. 136). The map suggests that the impacts in the southern and Pacific edges of the South America continent are beneficial. However, the marginal temperature impacts in the rest of Latin America are harmful. The impacts are especially harmful near the equator but they spread from northern Argentina to Mexico. Most of the inland continent will face a reduction of 100–250 USD per hectare. Figure 8.10 (see p. 137) presents a map of the marginal impact of precipitation on rainfed farms. The map suggests that the marginal impact of rainfall is harmful to most of Latin America and especially for Mexico. There is already substantial rain in a great deal of the continent, but some areas, such as Mexico, are dry. In Mexico, the harmful effect appears to be from winter precipitation (Mendelsohn et al., 2008). The impacts in the dry southern region of South America are much smaller.

Figure 8.11 (see p. 138) presents the marginal temperature effects for irrigated farms. The marginal temperature impacts on irrigated farms are similar to the rainfed impacts except more severe. The extent of the damage is larger. Virtually all regions will be damaged. The damage will be greatest near the Amazon basin and the equator. Figure 8.12 (see p. 139) presents the marginal precipitation effects for irrigated farms. The map suggests that the marginal impact of rainfall is positive for all regions in the Latin American continent, especially near the equator.

Overall, both irrigated and rainfed farms in Latin America are expected to react to climate change in a similar fashion but the effects on irrigated lands are more severe. Further, the response of farms in different regions in Latin America will be quite different. This range of results across the region has important policy implications which we turn to in Chapter 12.

The final analysis examines the impact of potential future climate scenarios on Latin American agriculture. The analysis examines the same two 2100 climate scenarios presented in the African simulation: PCM and CCC. As in the African case, the PCM scenario is mild and wet and the CCC scenario is hot and dry. The analysis examines only the effect of changes in climate. Other variables that are likely to change over time including prices, technology, capital and policies are not explored.

The climate simulations are presented in Table 8.9. With the mild climate scenario of PCM, there are losses to all types of farms. However, the predicted losses to small rainfed farms are expected to be small (4 percent) while the expected losses to small irrigated farms are larger (18 percent). Large irrigated and large rainfed losses are about the same (30 percent). With the CCC scenario, the losses to all farms are expected to

Table 8.9   Impacts from 2100 climate change scenarios in Latin America (USD per ha)

	Baseline	PCM Change	CCC Change
Small farms			
All	2088	–240**	–1016**
Rainfed	1890	–79	–662**
Irrigated	2617	–482*	–808**
Large farms			
All	1363	–266	–943**
Rainfed	1325	–395**	–1192**
Irrigated	1236	–370	–588

Notes:
One asterisk denotes significance at 5% level (*) and two asterisks at 1% level (**).
CCC: Canadian Climate Centre; PCM: Parallel Climate Model.
The estimates are changes in land value in USD per hectare of land calculated at the mean of the corresponding sample.

Source: Seo and Mendelsohn (2008c).

increase substantially. The losses to small rainfed farms rise to 35 percent and the losses to large rainfed farms rise to 31 percent. The losses to large irrigated farms rise to 48 percent and the losses to large irrigated farms rises to an astonishing 87 percent. In Latin America, all types of farms are vulnerable to warming, but surprisingly, irrigated farms are more vulnerable to warming than rainfed farms.

Figure 8.13 (see p. 140) presents a map of the impacts of the PCM climate scenario on small farms in South America. The results indicate that this scenario will be beneficial to small farms throughout South America but especially in the southern part of the continent. However, land values of small farms in Central America and especially near the equator are expected to drop, even with the mild PCM scenario. Figure 8.14 (see p. 141) presents the same map for large farms. The results for large farms are quite similar. There is little difference in how small and large farms react to the PCM scenario.

Figure 8.15 (see p. 142) presents a map of the impacts of the CCC climate scenario on Latin American small farms. The picture is dramatically different from Figure 8.13, with small farms in most of the continent suffering large damages. Only farms along the Pacific coast, the southern tier of South America, and in Central America actually benefit from this scenario. A similar map is shown for large farms in Figure 8.16 (see p. 143). The CCC climate scenario causes land values of large farms to fall in much the same places as with small farms. Large farms also benefit where small farms benefit.

CHINA

Our final analysis of individual farm-level data is from a study in China (Wang et al., 2009). The study is based on information collected across 45 000 households from the Household Income and Expenditure Survey (Chinese National Bureau of Statistics, 1995). The analysis examined all farmers in counties with meteorological stations. A total of 8405 farms across 124 counties and 28 provinces were examined. Although land value data was not available in the survey, the survey included information on farm size, farm type and net revenue. The farm data was then matched with climate and soil data. Similar to the African study, household labor could not be valued and so was omitted from net revenue.

Net revenue (measured in yuan per hectare) is then regressed on climate, soils and other control variables. The regression examines climate from all four seasons. The following control variables were also included: two soil variables, a plains dummy, road access, township access, irrigation, production association, education, land per household and altitude. The regression results are shown in Table 8.10. Separate regressions are run on all farms, irrigated farms, and rainfed farms. Note that the subsamples of rainfed and irrigated farms do not sum to the entire sample as it is not possible to determine whether some farms are irrigated or not. The farms that are hard to classify are left out of the subsamples. In the regression of all farms, six of the eight temperature coefficients are significant (only the autumn coefficients are not) and six of the eight precipitation coefficients are significant (only the summer coefficients are not). With the irrigated regression, half the temperature and all the precipitation coefficients are significant. Finally, with the rainfed subsample, six of the temperature and five of the precipitation coefficients are significant. The climate coefficients are significant.

Many of the control variables are also significant. Clay soil, silt soil, being on a plain, being located near a road, participating in a production association, all increase the value of farms in the complete sample. Larger farms have lower net revenues per ha. Again this could be due to the fact that household labor is not included as a cost in net revenue. Higher elevations reduce farm value as expected. Looking at irrigated farms, silt soils and participating in a production association increase net revenue. However, net revenue falls if the farm is on a plain or is larger. Finally, looking at just rainfed farms, farm net revenue increases on a plain or near a road, but it decreases in larger farms and at higher elevations.

Table 8.10   Chinese Ricardian regressions of net crop revenue

	Net crop revenue (Yuan/ha)		
	All farms	Irrigated	Rainfed
Spring temp.	1,453	4,149	1,789
	(2.18)*	(1.79)	(1.54)
Spring temp. sq.	–118.1	–170.4	–106.9
	(5.88)**	(2.18)*	(2.97)**
Summer temp.	–1,803	1,263	–6,200
	(2.01)*	(0.57)	(4.75)***
Summer temp. sq.	48.7	17.0	125.9
	(2.53)*	(0.35)	(4.03)***
Autumn temp.	119	–5,178	2,678
	(0.20)	(2.55)*	(2.54)*
Autumn temp. sq.	–12.1	67.7	–116.1
	(0.56)	(0.93)	(2.60)*
Winter temp.	1,226	2,064	911
	(4.44)**	(3.64)**	(1.66)
Winter temp. sq.	62.6	63.9	67.2
	(7.34)**	(2.91)*	(4.87)**
Spring prec.	–300.6	–268.3	–132.3
	(8.52)**	(2.84)*	(1.50)
Spring prec. sq.	1.057	0.726	0.605
	(8.56)**	(2.21)*	(1.69)
Summer prec.	5.61	151.1	–76.5
	(0.39)	(3.68)**	(2.70)*
Summer prec. sq.	–0.061	–0.241	0.132
	(1.55)	(2.22)*	(1.64)
Autumn prec.	–107.4	–413.8	–171.6
	(2.92)*	(3.67)**	(2.71)*
Autumn prec. sq.	0.944	2.311	1.276
	(5.31)**	(3.22)**	(4.25)**
Winter prec.	554.4	668.9	655.9
	(8.07)**	(3.43)**	(5.33)**
Winter prec. sq.	–6.355	–5.212	–8.248
	(7.96)**	(2.42)*	(5.27)**
Share of clay soil	4,360	201	–109
	(7.26)**	(0.14)	(0.08)
Share of silt soil	2,080	2,865	747
	(3.85)**	(2.68)**	(0.79)
Plain (1=Yes; 0=No)	856	–1,459	1,248
	(2.57)*	(1.96)*	(2.11)*
Road (1=Yes; 0=No)	2,022	722	3,313
	(2.96)**	(0.55)	(3.66)**

Table 8.10 (continued)

	Net crop revenue (Yuan/ha)		
	All farms	Irrigated	Rainfed
Distance to township	21.9	83.4	–35.8
government	(0.77)	(1.19)	(0.93)
Share of irrigation in village	4.6		
	(1.11)		
If participate production	1,713	2,941	–2,168
association			
(1=Yes; 0=No)	(2.50)*	(2.57)*	(1.27)
Share of labor without	4.9	24.6	–9.3
education	(0.71)	(1.71)	(0.90)
Log of cultivated land per	–5,189	–4,942	–3,934
household	(29.46)**	(13.72)**	(14.53)**
Altitude	–1.96	–0.92	–3.49
	(4.56)**	(1.41)	(2.46)*
Constant	26,242	–4,167	70,431
	(3.28)**	(0.19)	(5.22)**
Observations	8405	2750	2119
Adjusted R^2	0.21	0.17	0.26
F-test	89.23		

Notes:
Dependent variable is yuan/ha.
One asterisk (*) denotes significance at 5% level and two asterisks (**) denotes significance at 1% level.
F-test measures whether coefficients of rainfed and irrigated farmland are significantly different from combined regression.

Source: Wang et al. (2008, Table 2).

The marginal effect of climate on net revenue per year is calculated in Table 8.11. Values are converted from yuan to USD to make them comparable with Tables 8.2, 8.5 and 8.8. The results reveal that the marginal temperature effect is negative but small in the full sample. Looking at the subsamples, there is a positive effect from temperature on irrigated land and a large negative effect on rainfed land. Since the total amount of land is almost evenly split in China between rainfed and irrigated land, the two effects are cancelling each other out in the full sample. Although some parts of China are hot, the average temperature is temperate. Combined with the large reliance on irrigated farms, China's crop sector is not that sensitive to marginal changes in temperature. The effect of increased precipitation is positive in all three regressions. Rain increases the value of irrigated and rainfed farms alike.

Table 8.11   Marginal impacts of climate on crop net revenue in China

	All farms	Irrigated farms	Rainfed farms
Temperature (USD/ha/°C)			
Annual	–10*	68*	–95*
Annual Elasticity	–0.09*	0.62*	–0.88*
Precipitation (USD/ha/mm/mo)			
Annual	15*	27**	23*
Annual Elasticity	0.80*	1.48**	1.24*

Notes:
Yuan converted to 2006 USD using exchange rate of 8 Yuan/USD.
One asterisk denotes significance at 5% level (*) and two asterisks denotes significance at 1% level (**).
Values calculated at mean climate.

Source: Wang et al. (2008).

The marginal effects of temperature on Chinese rainfed farms are mapped in Figure 8.17 (see p. 144). The map reveals that rainfed farms in the southern tier of China are very sensitive to increased temperature and would suffer large reductions in net revenue. These marginal damages would extend to most of China's productive rainfed lands. Only the most northern and western lands would benefit. Figure 8.18 (see p. 145) presents the marginal precipitation effects for rainfed farms. Increased rainfall would be mildly beneficial except in China's wet southeast region where it would cause a small damage. The results for China's irrigated land are quite different. Figure 8.19 (see p. 146) shows that increased marginal temperatures would increase net revenues of irrigated land especially in the southeast and southwest regions. In the southeast, warmer temperatures extend the growing season and make double cropping possible. In the southwest, warmer temperatures would simply make irrigation more profitable. There would also be small benefits in central China. Only irrigated lands in the north would do less well in a warmer world. Figure 8.20 (see p. 147) presents the results for precipitation. Net revenues of irrigated farms would increase with more rain almost uniformly across the country.

The analysis was not able to run a detailed PCM and CCC scenario for China that allows different changes in each province. However, in order to get a sense of the consequences of mild and harsh climate scenarios, two uniform climate scenarios are explored in Table 8.12. One scenario increases the temperature by 2.5°C and precipitation by 8 percent. The second scenario increases temperature by 5°C and does not increase precipitation at all. The mild scenario suggests that there will be agricultural

Table 8.12   Uniform climate scenario impact on China (USD per ha)

	Baseline	+2.5°C + 8% Precip	+5°C
		(Change)	(Change)
All	1268	–75.3	–719.4
Rainfed	1540	–262.3	–2083.2
Irrigated	933	79.6	18.4

Source: Wang et al. (2009).

losses of about 6 percent overall. However, irrigated farms will actually increase in value slightly, while rainfed farms fall in value about 17 percent. With the harsh scenario, the overall analysis suggests that agricultural net revenues will fall by about 57 percent, which is a substantial loss for China. Irrigated farms will be hardly affected provided they have sufficient water. However, rainfed farms are predicted to become unprofitable at the mean. Obviously, some cooler places will still be able to support rainfed agriculture but this will be a major blow to China's agriculture. If water supplies fall and farmers need to switch from irrigated to rainfed agriculture, the damages will be even higher.

CONCLUSION

The individual farm analyses have been able to shed light across the regions of the world most vulnerable to climate change. In the regions near the tropics, where farming is at greatest peril from warming, these studies have revealed that even with adaptation, climate change may lead to serious reductions in net revenues. The research has not yet covered every country in the world that is at risk. But it has covered enough of the terrain that experts can now make informed judgments about the potential impacts of climate change on agriculture.

Interestingly, the climate models in each continent tested so far have been different. For example, the South American farms are the most sensitive to climate change. Within South American agriculture, irrigated farms are much more sensitive to climate change compared to rainfed farms. Losses to irrigated farms could be addressed by proper policy, as will be discussed in Chapter 12. Despite their low productivity, African farms are less sensitive to climate change than South American farms. Because of their greater reliance on irrigation, farms in China appear to be resilient to at least small changes in climate.

There are some additional advantages of relying on individual farm data compared to at least current aggregate data. One can use individual farm data to separate the contribution of irrigated versus rainfed farms. The data can also be used to test whether small versus large farms have similar climate sensitivity. The data on net revenues can further be used to separate the contribution of livestock versus crops. Each of these comparisons has been quite insightful. First, as expected from the research first presented by Schlenker et al. (2005), except in South America, irrigated crop farms are less sensitive to climate than rainfed crop farms. Second, small farms are often less sensitive to climate than large farms. Although large farms are more commercial and therefore have better access to capital and technology, they are also more specialized. Small farms consequently have more alternatives as climates warm or dry. Third, livestock are less sensitive to climate than crops. Farmers can therefore turn to livestock as an alternative if crops become marginalized. Mixed farms that have both crops and livestock are therefore more robust against climate change than more specialized farms.

REFERENCES

Boer, G., G. Flato and D. Ramsden (2000), 'A transient climate change simulation with greenhouse gas and aerosol forcing: projected climate for the 21st century', *Climate Dynamics*, **16**, 427–50.

Chinese National Bureau of Statistics (1995), 'Household income and expenditure survey', Beijing, China.

Dinar, A., R. Hassan, R. Mendelsohn and J. Benhin (2008), *Climate Change and Agriculture in Africa: Impact Assessment and Adaptation Strategies*, London, UK: Earthscan.

Fleischer, A., I. Lichtman and R. Mendelsohn (2008), 'Climate change, irrigation, and Israeli agriculture: will warming be harmful?', *Ecological Economics*, **65**, 508–15.

IPCC (Intergovernmental Panel on Climate Change) (2007), *State of the Science*, Cambridge, UK: Cambridge University Press.

Kurukulasuriya, P. and R. Mendelsohn (2008), 'A Ricardian analysis of the impact of climate change on African cropland', *African Journal of Agriculture and Resource Economics*, **2**, 1–23.

Kurukulasuriya, P., R. Mendelsohn, R. Hassan, J. Benhin, M. Diop, H. M. Eid, K.Y. Fosu, G. Gbetibouo, S. Jain, A. Mahamadou, S. El-Marsafawy, S. Ouda, M. Ouedraogo, I. Sène, S.N. Seo, D. Maddison and A. Dinar (2006), 'Will African agriculture survive climate change?', *World Bank Economic Review*, **20**, 367–88.

Mendelsohn, R., A.F. Avila and S.N. Seo (2007), *Incorporation of Climate Change into Rural Development Strategies: Synthesis of the Latin American Results*, Montevideo, Uruguay: PROCISUR.

Seo, N.S. and R. Mendelsohn (2006), 'Climate change adaptation in Africa: a microeconomic analysis of livestock choice', World Bank Policy Research Working Paper 4277, Washington, DC, USA: World Bank.

Seo, S.N. and R. Mendelsohn (2008a), 'Climate change impacts and adaptations on animal husbandry in Africa', *African Journal of Agriculture and Resource Economics*, **2**, 65–82.

Seo, S.N. and R. Mendelsohn (2008b), 'Measuring impacts and adaptation to climate change: a structural Ricardian model of African livestock management', *Agricultural Economics*, **38**, 150–65.

Seo, S.N. and R. Mendelsohn (2008c), 'A Ricardian analysis of the impact of climate change on Latin American farms', *Chilean Journal of Agricultural Research*, **68**(1), 69–79.

Schlenker, W., W.M. Hanemann and A.C. Fisher (2005), 'Will US agriculture really benefit from global warming? Accounting for irrigation in the hedonic approach', *American Economic Review*, **95**, 395–406.

Wang, J., R. Mendelsohn, A. Dinar, J. Huang, S. Rozelle and L. Zhang (2009), 'The impact of climate change on China's agriculture', *Agricultural Economics*, **40**, 323–37.

Washington, W., J. Weatherly, G. Meehl, A. Semtner, T. Bettge, A. Craig, W. Strand, J. Arblaster, V. Wayland, R. James and Y. Zhang (2000), 'Parallel Climate Model (PCM): control and transient scenarios', *Climate Dynamics*, **16**, 755–74.

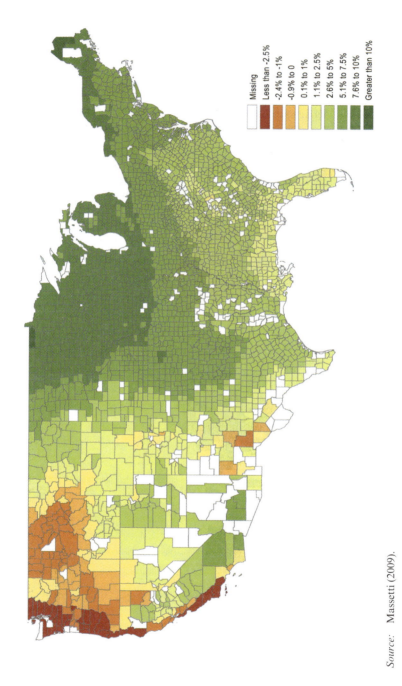

Source: Massetti (2009).

Figure 7.1 Impact of marginal temperature increase on United States farmland values

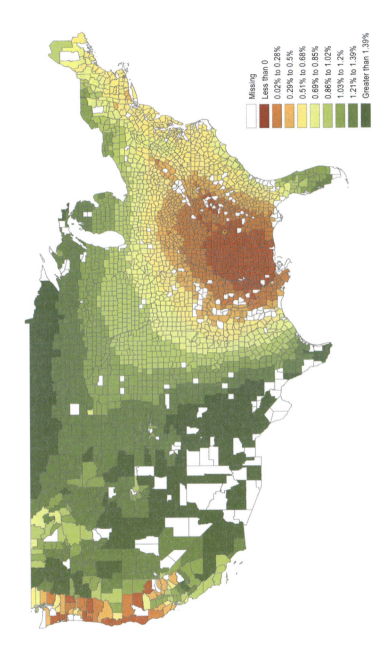

Source: Massetti (2009).

Figure 7.2   Impact of marginal precipitation increase on United States farmland values

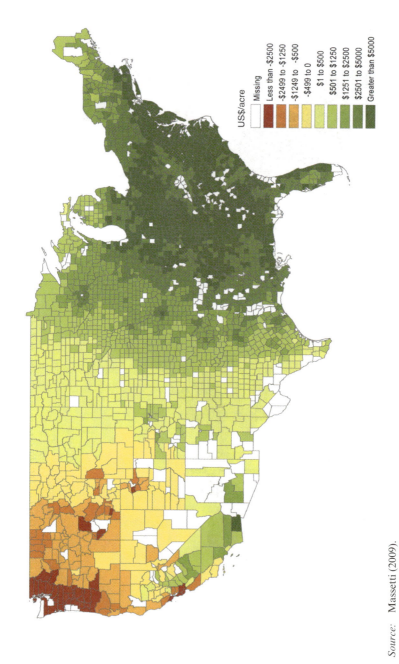

US$/acre	
	Missing
	Less than -$2500
	-$2499 to -$1250
	-$1249 to -$500
	-$499 to 0
	$1 to $500
	$501 to $1250
	$1251 to $2500
	$2501 to $5000
	Greater than $5000

Source: Massetti (2009).

Figure 7.3   Impact of PCM climate scenario on United States farmland values

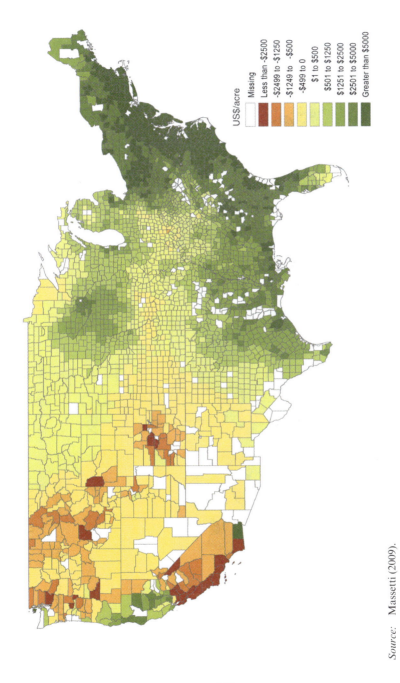

US$/acre

Missing
Less than -$2500
-$2499 to -$1250
-$1249 to -$500
-$499 to 0
$1 to $500
$501 to $1250
$1251 to $2500
$2501 to $5000
Greater than $5000

Source: Massetti (2009).

Figure 7.4 Impact of Hadley III climate scenario on United States farmland values

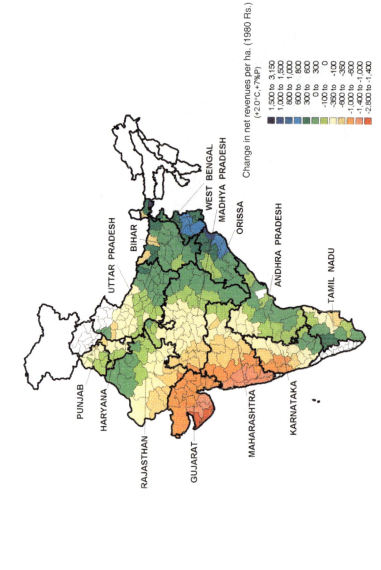

Change in net revenues per ha. (1980 Rs.)
(+2.0°C, +7%P)

1,500 to 3,150
1,000 to 1,500
800 to 1,000
600 to 800
300 to 600
0 to 300
-100 to 0
-350 to -100
-600 to -350
-1,000 to -600
-1,400 to -1,000
-2,800 to -1,400

Source: Sanghi and Mendelsohn (2008).

Figure 7.5   Impact of a 2.0°C warming on crop net revenue in India

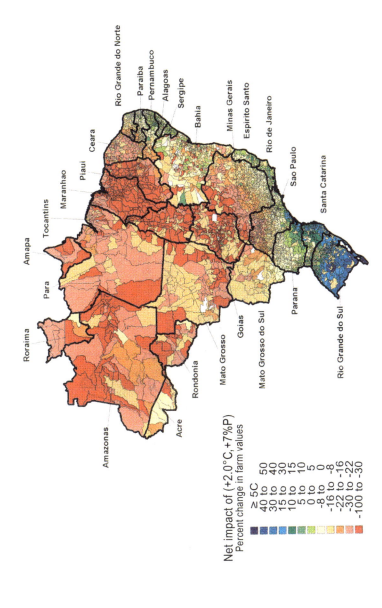

Net impact of (+2.0°C,+7%P)
Percent change in farm values

| ≥ 5C |
40 to	50
30 to	40
15 to	30
10 to	15
5 to	10
0 to	5
-8 to	0
-16 to	-8
-22 to	-16
-30 to	-22
-100 to	-30

Source: Sanghi and Mendelsohn (2008).

Figure 7.6 Impact of a 2.0°C warming on farmland value in Brazil

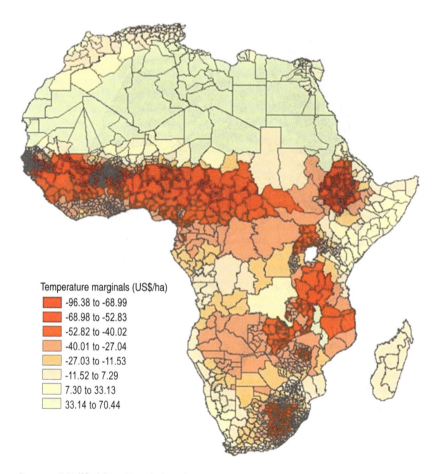

Temperature marginals (US$/ha)

- ■ -96.38 to -68.99
- ■ -68.98 to -52.83
- ■ -52.82 to -40.02
- ■ -40.01 to -27.04
- ■ -27.03 to -11.53
- ■ -11.52 to 7.29
- ■ 7.30 to 33.13
- ■ 33.14 to 70.44

Source: Modified from Kurukulasuriya et al. (2006).

Figure 8.1 Marginal temperature impact on crop net revenues from African farms

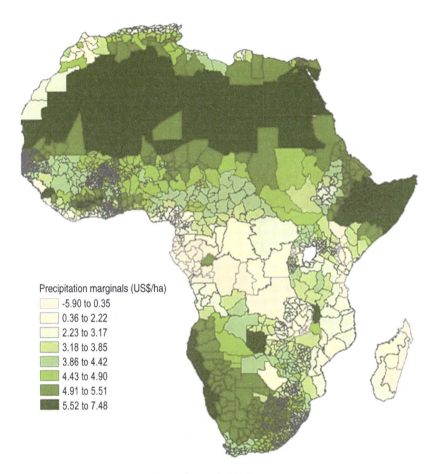

Precipitation marginals (US$/ha)

- -5.90 to 0.35
- 0.36 to 2.22
- 2.23 to 3.17
- 3.18 to 3.85
- 3.86 to 4.42
- 4.43 to 4.90
- 4.91 to 5.51
- 5.52 to 7.48

Source: Modified from Kurukulasuriya et al. (2006).

Figure 8.2 Marginal precipitation impact on crop net revenues from African farms

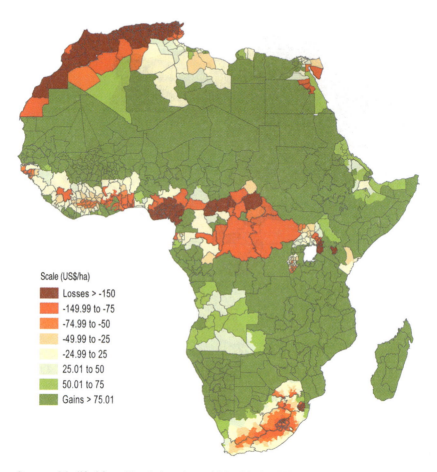

Scale (US$/ha)

■ (dark red)	Losses > -150
■ (orange-red)	-149.99 to -75
■ (orange)	-74.99 to -50
■ (tan)	-49.99 to -25
■ (cream)	-24.99 to 25
■ (pale green)	25.01 to 50
■ (light green)	50.01 to 75
■ (green)	Gains > 75.01

Source: Modified from Kurukulasuriya and Mendelsohn (2007).

Figure 8.3   Impact of a PCM 2100 climate scenario on crop net revenues from African farms

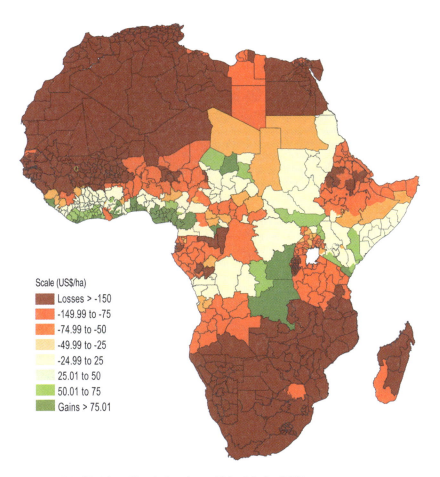

Scale (US$/ha)

■	Losses > -150
■	-149.99 to -75
■	-74.99 to -50
■	-49.99 to -25
■	-24.99 to 25
■	25.01 to 50
■	50.01 to 75
■	Gains > 75.01

Source: Modified from Kurukulasuriya and Mendelsohn (2007).

Figure 8.4   *Impact of a CCC 2100 climate scenario on crop net revenue from African farms*

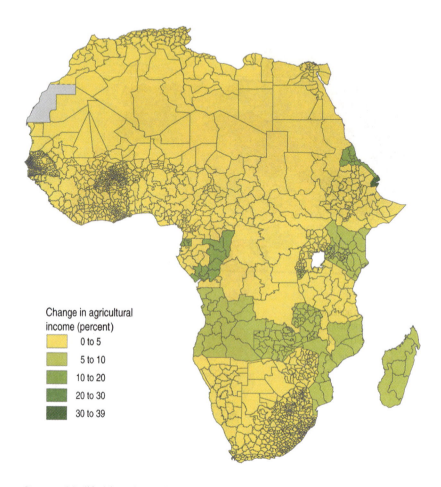

Source: Modified from Seo and Mendelsohn (2007a).

Figure 8.5   Impact of a PCM 2100 climate scenario on livestock net revenues from small African farms

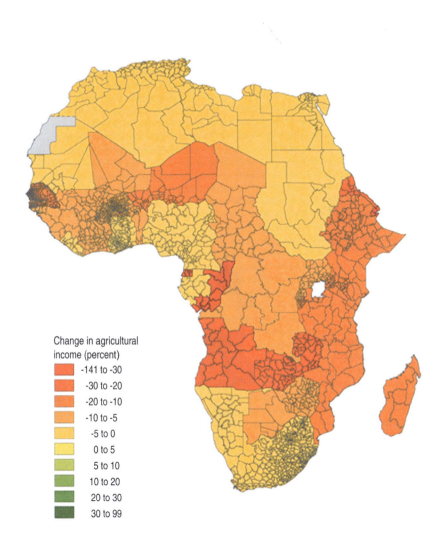

Change in agricultural
income (percent)
-141 to -30
-30 to -20
-20 to -10
-10 to -5
-5 to 0
0 to 5
5 to 10
10 to 20
20 to 30
30 to 99

Source: Modified from Seo and Mendelsohn (2007a).

Figure 8.6 *Impact of a PCM 2100 climate scenario on livestock net*
revenues from large African farms

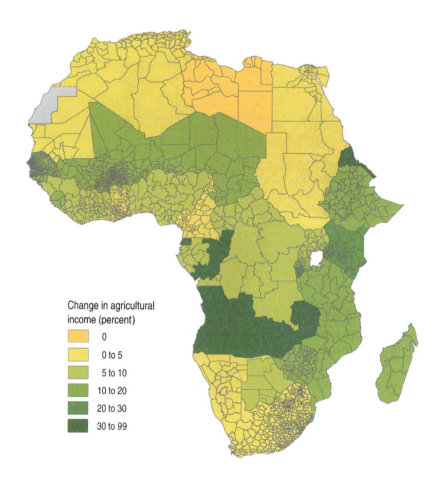

Change in agricultural income (percent)

- 0
- 0 to 5
- 5 to 10
- 10 to 20
- 20 to 30
- 30 to 99

Source: Modified from Seo and Mendelsohn (2007a).

*Figure 8.7   Impact of a CCC 2100 climate scenario on livestock net
revenues from small African farms*

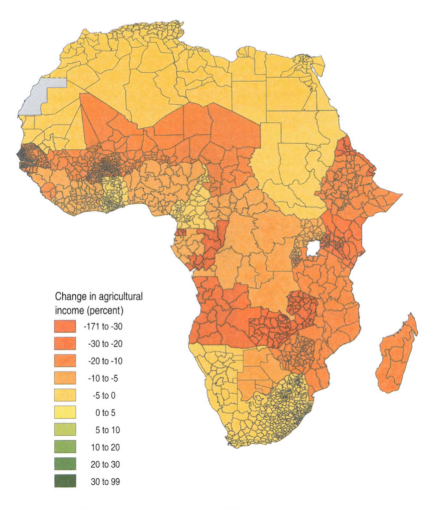

Change in agricultural
income (percent)

- -171 to -30
- -30 to -20
- -20 to -10
- -10 to -5
- -5 to 0
- 0 to 5
- 5 to 10
- 10 to 20
- 20 to 30
- 30 to 99

Source: Modified from Seo and Mendelsohn (2007a).

*Figure 8.8   Impact of a CCC 2100 climate scenario on livestock net
revenues from large African farms*

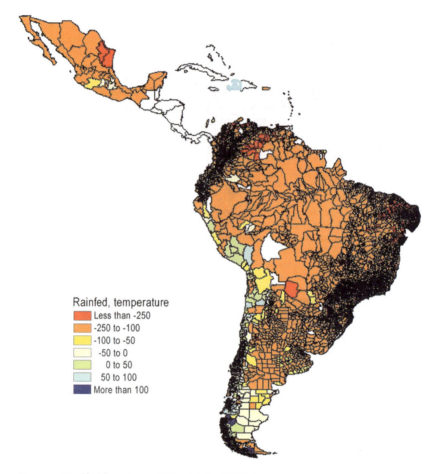

Source: Modified from Seo and Mendelsohn (2008c).

Figure 8.9   *Marginal temperature impact on farmland values from Latin American rainfed farms*

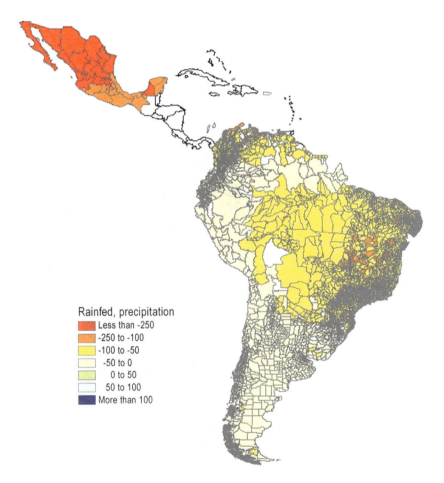

Rainfed, precipitation

■	Less than -250
■	-250 to -100
■	-100 to -50
■	-50 to 0
■	0 to 50
■	50 to 100
■	More than 100

Source: Modified from Seo and Mendelsohn (2008c).

Figure 8.10   Marginal precipitation impact on farmland values from Latin American rainfed farms

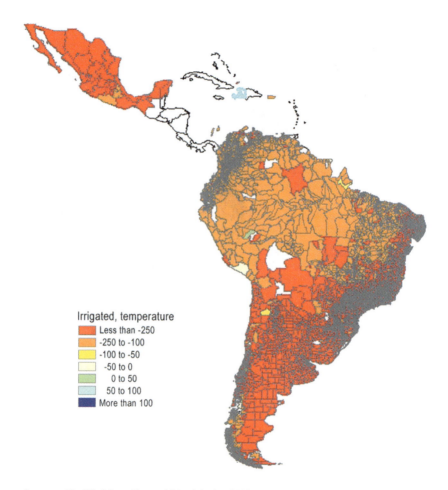

Irrigated, temperature
- Less than -250
- -250 to -100
- -100 to -50
- -50 to 0
- 0 to 50
- 50 to 100
- More than 100

Source: Modified from Seo and Mendelsohn (2008c).

Figure 8.11   Marginal temperature impact on farmland values from Latin American irrigated farms

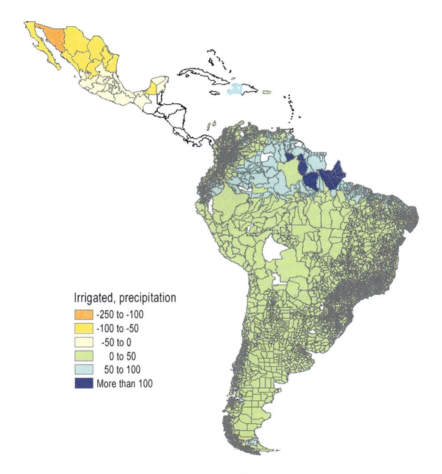

Source: Modified from Seo and Mendelsohn (2008c).

Figure 8.12   Marginal precipitation impact on farmland values from Latin American irrigated farms

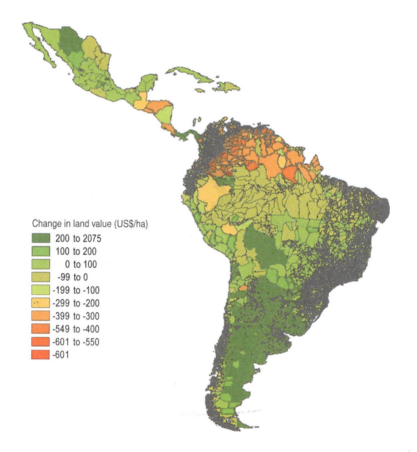

Change in land value (US$/ha)
- 200 to 2075
- 100 to 200
- 0 to 100
- -99 to 0
- -199 to -100
- -299 to -200
- -399 to -300
- -549 to -400
- -601 to -550
- -601

Source: Modified from Seo and Mendelsohn (2008c).

*Figure 8.13   Impact of a PCM 2100 climate scenario on farmland values
of small farms in Latin America*

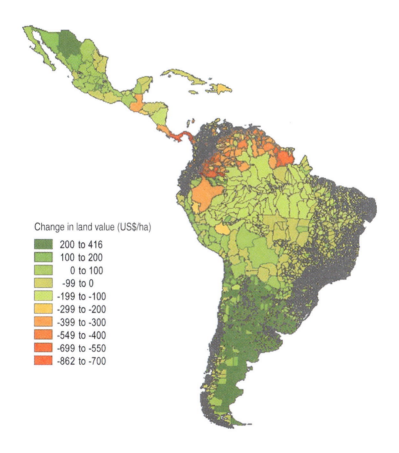

Change in land value (US$/ha)

■	200 to 416
■	100 to 200
■	0 to 100
■	-99 to 0
■	-199 to -100
■	-299 to -200
■	-399 to -300
■	-549 to -400
■	-699 to -550
■	-862 to -700

Source: Modified from Seo and Mendelsohn (2008c).

Figure 8.14 Impact of a PCM 2100 climate scenario on farmland values of large farms in Latin America

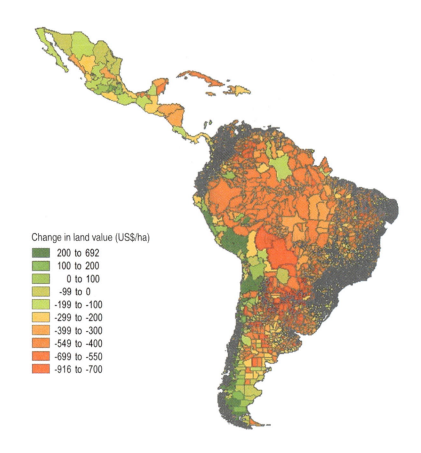

Change in land value (US$/ha)

▮	200 to 692
▮	100 to 200
▮	0 to 100
▮	-99 to 0
▮	-199 to -100
▮	-299 to -200
▮	-399 to -300
▮	-549 to -400
▮	-699 to -550
▮	-916 to -700

Source: Modified from Seo and Mendelsohn (2008c).

*Figure 8.15   Impact of a CCC climate scenario on farmland values of
            small farms in Latin America*

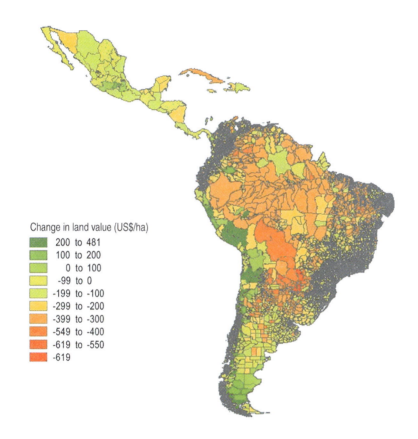

Change in land value (US$/ha)
- 200 to 481
- 100 to 200
- 0 to 100
- -99 to 0
- -199 to -100
- -299 to -200
- -399 to -300
- -549 to -400
- -619 to -550
- -619

Source: Modified from Seo and Mendelsohn (2008c).

Figure 8.16 Impact of a CCC 2100 climate scenario on farmland values of large farms in Latin America

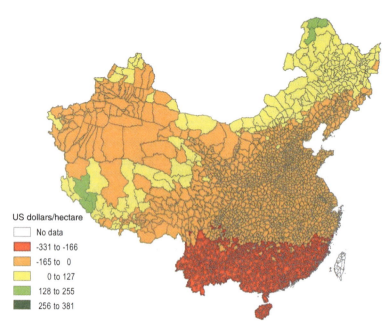

US dollars/hectare

☐ No data
🟥 -331 to -166
🟧 -165 to 0
🟨 0 to 127
🟩 128 to 255
🟩 256 to 381

Source: Modified from Wang et al. (2008).

Figure 8.17   Marginal temperature impact on crop net revenues from China's rainfed farms

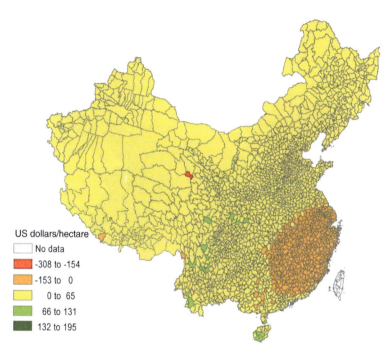

US dollars/hectare

☐ No data
■ -308 to -154
■ -153 to   0
■ 0 to 65
■ 66 to 131
■ 132 to 195

Source: Modified from Wang et al. (2008).

Figure 8.18   Marginal precipitation impact on crop net revenues from China's rainfed farms

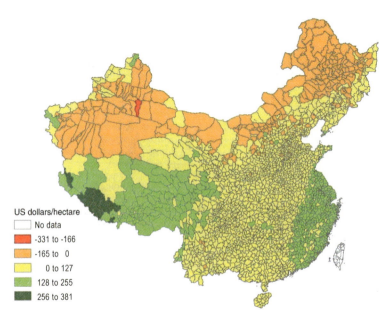

US dollars/hectare
☐ No data
■ -331 to -166
■ -165 to 0
■ 0 to 127
■ 128 to 255
■ 256 to 381

Source: Modified from Wang et al. (2008).

Figure 8.19   Marginal temperature impact on crop net revenues from China's irrigated farms

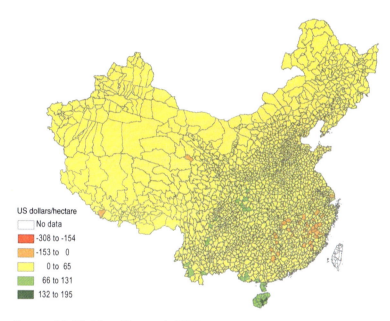

US dollars/hectare
- No data
- -308 to -154
- -153 to 0
- 0 to 65
- 66 to 131
- 132 to 195

Source: Modified from Wang et al. (2008).

Figure 8.20   *Marginal precipitation impact on crop net revenues from China's irrigated farms*

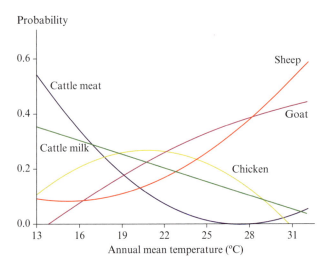

Note: Mean temperature: cattle meat=19; cattle milk=19; goat=24; sheep=24; chicken=21.

Source: Modified from Seo and Mendelsohn (2008d).

Figure 9.1   *Effect of temperature on the probability of choosing African livestock*

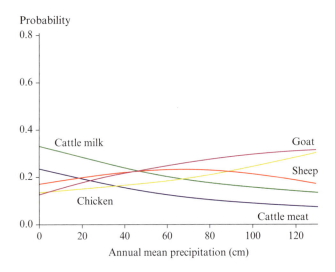

Note: Mean precipitation: cattle meat=58; cattle milk=63; goat=68; sheep=59; chicken=76.

Source: Modified from Seo and Mendelsohn (2008c).

Figure 9.2   *Effect of precipitation on the probability of choosing African livestock*

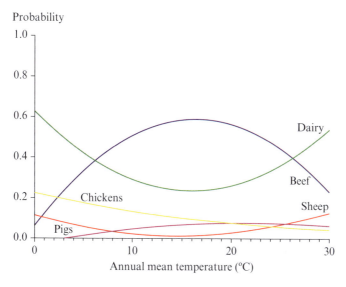

Source: Modified from Seo and Mendelsohn (2007d).

Figure 9.3 Effect of temperature on choice of Latin American livestock

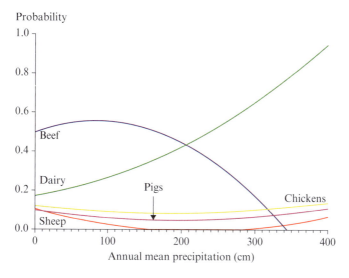

Source: Modified from Seo and Mendelsohn (2007d).

Figure 9.4 Effect of precipitation on choice of Latin American livestock

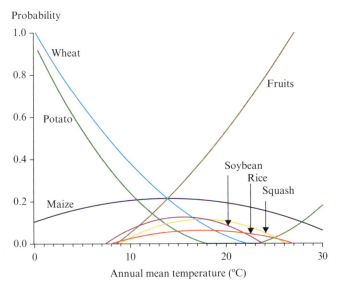

Source: Modified from Seo and Mendelsohn (2007c).

Figure 9.5   Effect of temperature on choice of Latin American crops

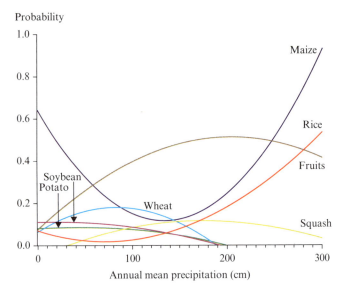

Source: Modified from Seo and Mendelsohn (2007c).

Figure 9.6   Effect of precipitation on choice of Latin American crops

9. Adaptation studies

We reviewed the general literature on climate adaptation in Chapter 5. In this review, we examined the literature describing theory, case studies and quantitative approaches to measuring climate adaptation. In this chapter, we examine the empirical results of several quantitative studies that rely on cross-sectional evidence to measure how farmers have responded to climate. The quantitative studies use statistical modeling to capture general behavior across broad populations.

There have been several quantitative studies of adaptation to climate change. Some studies of US farms have calculated shifts in crops using mathematical programming (Adams et al., 1990, 1999). Some mathematical programming models have also been used to allocate water across users. These studies, in for example, California, the Colorado River basin and the Nile River basin, have found that it is often important to allocate water from low valued irrigation projects to higher valued uses (Howitt and Pienaar, 2006; Hurd et al., 1999; Lund et al., 2006; Strzepek et al., 1996). There have also been some programming models of individual farms to capture the adjustment costs of climate change (Kaiser et al., 1993; Kelly et al., 2005).

The focus of this chapter is upon quantitative analysis of cross-sectional data. The advantage of cross-sectional data is that it measures long-run adaptation behavior to the climate of each farm. By matching farms to climates, one can examine how farm decisions vary by climate. Endogenous choices by farmers to own livestock, choose crop types, pick livestock species, determine herd size, and install irrigation can all be examined with cross-sectional analysis. The standing hypothesis is that these choices are sensitive to climate.

AFRICAN STUDIES

In Chapter 5, we discussed the use of a choice model to better understand adaptation decisions. We now apply this choice model to examine a number of different adaptations that farmers make in Africa. The first decision we explore is whether or not African farmers choose livestock (Seo and Mendelsohn, 2008d). We rely on a sample of farmers selected across 10

Table 9.1   African livestock decision

Variable	Coefficient	P value
Intercept	1.349	0.09
Summer temperature	0.0158	0.78
Summer temp. sq.	−0.00052	0.64
Summer precipitation	−0.00385	0.03
Summer prec. sq.	8.64E-06	0.16
Winter temperature	0.00918	0.89
Winter temp. sq.	0.000156	0.93
Winter precipitation	0.00443	0.12
Winter prec. sq.	−0.00007	<0.0001
Flow	0.000496	0.94
Head farm	0.104	0.44
Electricity	0.165	0.03
Soil Ferralsols	1.71	0.34
Soil Luvisols	1.12	0.02
Soil Vertisols	−0.29	0.72
Milk price, kg	−0.00103	0.00

Note: Dependent variable is whether or not to own livestock.

Source: Seo and Mendelsohn (2008d).

countries: Burkina Faso, Cameroon, Egypt, Ethiopia, Ghana, Kenya, Niger, Senegal, South Africa and Zambia. The countries were selected to reflect a considerable variation of climate across countries and the samples within each country were selected to get variation within countries as well. The farmers represent both small and large operations; household farms as well as commercial farms.

Table 9.1 presents a logit analysis estimated across over 8000 farms concerning the choice over whether or not to own livestock. The results indicate that livestock ownership is sensitive to climate. Both summer and winter precipitation influence whether livestock are chosen. Curiously, temperature does not have a significant effect. Wetter climates reduce livestock ownership. This is partly because farmers switch to crops in wetter settings. However, it is also true that the African landscape shifts from grasslands to forests as precipitation increases, making them less attractive for forage. Finally, warm, wet settings tend to have more livestock diseases. Other variables also influence livestock ownership. Electricity and Luvisol soils increase the chance of ownership and higher milk prices reduce ownership.

A more complex choice facing livestock owners is which species to select as their primary source of net revenue. This analysis is only conducted

Table 9.2   African livestock species choice relative to chickens

	Beef cattle		Goats		Dairy cattle		Sheep	
	Coefficient	chi-sq	Coefficient	chi-sq	Coefficient	chi-sq	Coefficient	chi-sq
Intercept	2.32	0.68	0.31	0.02	−1.46	0.6	3.76	3.14
Temp. summer	0.33	2.59	−0.38	6.78	−0.39	10.61	−0.37	6.89
Temp. summer sq.	−0.005	1.59	0.008	8.75	0.008	10.74	0.006	5.33
Temp. winter	−1.04	49.1	0.30	2.62	0.3	5.68	0.03	0.03
Temp. winter sq.	0.026	38.43	−0.004	0.83	−0.005	2.32	0.005	1.41
Prec. summer	0.003	0.27	−0.012	6.83	−0.01	8.37	−0.017	13.18
Prec. summer sq.	4.62E-06	0.04	0.00005	12.08	0.000024	2.31	0.00004	4.8
Prec. winter	0.025	7.67	−0.02	8.15	0.04	41.71	−0.02	12.17
Prec. winter sq.	−0.00009	2.44	0.00008	4.5	−0.0002	39.43	0.00001	0.11
Cambisols	1.32	1.54	1.08	2.9	2.65	16.7	1.58	6.57
Gleysols	0.95	0.33	−2.87	4.71	−3.9	7.44	−3.72	5.75
Luvisols	0.69	4.78	0.21	1.03	−0.2	0.48	0.38	3.34
Vertisols	−0.15	0.02	1.02	1.98	−3.84	3.36	1.03	2.02
Planasols	1.32	1.19	−2.58	7.14	2.15	4.29	−3.46	11.54
Flow	0.011	0.28	−0.028	0.9	0.09	32.59	−0.02	0.83
Electricity	−0.42	14.46	−0.08	1	0.14	2.51	−0.28	12.51
Beef price	0.02	0.01	−0.02	0.02	0.22	3.91	0.16	1.38
Milk price	−0.39	7.41	0.07	0.42	0.37	12.71	0.037	0.12
Sheep price	0.003	0.58	−0.01	5.54	0.0005	0.02	−0.006	2.29
Eggs price	0.51	18.54	−0.08	1.05	1.23	237.74	0.03	0.21

Notes: The number of observations is 3693.
Likelihood ratio test: $P<0.0001$; Lagrange multiplier test: $P<0.0001$; Wald test: $P<0.0001$.

Source: Seo and Mendelsohn (2008d).

across farms with livestock. The analysis is further limited to farms that reported species choice (3693 farms). In Africa, farmers choose among five primary animals: beef cattle, dairy cattle, sheep, goats and chickens (Seo and Mendelsohn, 2008d). These primary animals generate 90 percent of the livestock income in the sample. A multinomial logit model is estimated to calculate the probability of each species relative to choosing chickens. The independent variables include winter and summer climate variables as well as five soil variables: electricity, surface flow, and beef, milk, sheep and egg prices.

The results in Table 9.2 indicate that farms with electricity choose dairy cattle more often, but choose beef cattle and sheep less often. Electricity is needed for dairy cattle farms to produce and store milk. Electricity may be endogenous – it may be delivered to farms that need it for production, so one must be careful interpreting this coefficient. When there is more surface water flow in a district, farmers are more likely to choose dairy cattle. Luvisol soils increase the choice of beef cattle but reduce the choice of dairy cattle. Gleysols and Vertisols reduce the choice of dairy cattle, but Cambisols and Planosols increase the likelihood a farmer will choose dairy cattle. As expected, when the milk price is higher, farmers choose goats and dairy cattle more often, but beef cattle less often. Surprisingly, when egg prices are higher, farmers are more likely to choose beef cattle and dairy cattle relative to chickens. In this case, it could well be that species choice is determining egg prices rather than egg prices determining species choice.

The next analysis explores the choice of herd size (Seo and Mendelsohn, 2008d). The herd size choice is conditional on choosing the species. Table 9.3 reports the five herd size regressions with selection bias correction terms from Table 9.3. The selection bias parameter estimates are significant and substantial. When a farmer is expected to choose beef cattle, the farmer tends to own a smaller number of dairy cattle but a larger number of goats and chickens. When the farmer is expected to choose chickens, the farmer tends to own a smaller number of goats and sheep. Other control variables are also significant. When the farmer has electricity, he/she owns more beef cattle and chickens but fewer sheep. When water flow is higher in a district, the size of the sheep operation tends to be smaller. With Cambisol soils, sheep operations are larger, while they are smaller under Planosol soils. When the dominant soil is Luvisols, farmers own more dairy cattle and goats. When it is Vertisols, they own fewer beef cattle. The key observation in Table 9.3, however, is that herd size is climate-sensitive. Climate variables affect the number of animals owned by the farms. Beef cattle are sensitive to summer precipitation and winter temperature. Sheep ownership is sensitive to summer precipitation, while chicken ownership is sensitive to winter precipitation.

To understand how climate change affects livestock species choice and herd size, we calculate the marginal effects of a temperature increase and a precipitation increase on the choice of each species and the number of that species at the mean climates in Africa in Table 9.4. If temperature increases by 1°C, the probability of choosing goats or sheep increases, while the probability of choosing beef cattle, dairy cattle or chickens decreases. Similarly, farmers will move away from dairy cattle and sheep towards goats and chickens if precipitation increases by 1 mm/mo. As temperatures increase, the number of chickens and especially beef cattle decreases but the number of goats, sheep and dairy cattle slightly increases. The effect of the reduction in beef cattle herd size is large in economic terms because

Table 9.3   African herd size

	Beef	Dairy	Goats	Sheep	Chickens
Intercept	531.4*	−22.28	45.7	57.1**	927
Temp. summer	−26.9	−0.24	−0.50	0.89	−39.1
Temp. summer sq.	0.61	0.009	0.01	−0.02	1.32
Prec. summer	−52.3*	2.06	−4.52	−8.4**	−94.3
Prec. summer sq.	1.21	−0.024	0.13*	0.23**	1.15
Temp. winter	3.09**	0.052	−0.07	−0.01	−0.56
Temp. winter sq.	−0.011**	−0.0004**	0.0002	−0.0001	0.02**
Prec. winter	−0.96	0.058	−0.11	−0.006	9.67**
Prec. winter sq.	0.01*	−0.0006*	0.0004	−7E-05	−0.03**
Cambisols	14.18	3.27	7.38	10.2**	39.4
Gleysols	−224.03	9.49	−29.01	−22.7	229.1
Luvisols	11.65	12.12**	8.17*	3.15	98.7
Vertisols	−170.91**	−25.5	29.3*	0.57	−140.3
Planosols	−100.2	−15.8	−8.29	−37.8**	838
Flow	−3.908	0.054	−0.42	−1.16**	−4.8
Electricity	73.7**	1.36	1.38	−3.03*	318.2**
Select beef		−24.7*	23.9**	10.8	867**
Select milk	9.64		−8.49	9.34**	41.1
Select goats	104.5	−1.9		−9.57	128
Select sheep	69.1	21.08	19.9		−1034*
Select chickens	−220.6	4.24	−34.4**	−19.5*	
N	323	1036	720	784	830
Adjusted R^2	0.31	0.09	0.05	0.05	0.14

Note: One asterisk (*) denotes significance at 5% level and two asterisks (**) denotes significance at 1% level.

Source: Seo and Mendelsohn (2008d).

of the relatively high value of each animal. More precipitation increases the number of chickens and beef cattle but decreases the number of goats, sheep and dairy cattle.

These results can be displayed in graphical terms. Figure 9.1 (see p. 148) shows how the probability of choosing each African species changes with temperature. For example, the graph shows that beef and dairy cattle are more likely to be chosen in cooler temperatures. However, in temperatures closer to the mean for Africa, chickens are more likely. Finally, as temperatures become even warmer, sheep and goats become more likely. Figure 9.2 (see p. 148) shows a similar graph for precipitation. In dry conditions, beef and dairy cattle are slightly preferred. However, in wetter conditions, farmers shift to goats and chickens. These results may well reflect

Table 9.4   Marginal changes from climate

	Beef	Dairy	Goats	Sheep	Chickens
Species choice:					
Baseline	6.26%	28.30%	19.84%	23.46%	22.14%
Temperature	−0.31%	−0.02%	+0.51%	+1.75%	−1.94%
Precipitation	+0.05%	−0.04%	+0.11%	−0.14%	+0.03%
Herd size:					
Baseline	49.89	5.59	11.46	12.56	108.88
Temperature	−10.04	+1.33	+2.07	+2.18	−25.20
Precipitation	+0.30	−0.06	−0.02	−0.03	+6.58

Source: Seo and Mendelsohn (2008d).

the long-term ecosystem response to climate. For example, with less precipitation, the ecosystem is more likely to be savannah with more grazing opportunities for cattle and sheep. However, with a wetter climate, the ecosystem is more likely to be forested, which favors chickens and goats.

The next analysis explores whether climate affects crop choice in Africa (Kurukulasuriya and Mendelsohn, 2008a). The choice set includes three different crops and six crop combinations: maize, cowpea, sorghum, fruit–vegetables, maize–beans, cowpea–sorghum, maize–groundnut, maize–millet, and millet–groundnut. A multinomial regression is estimated across over 5000 farms. The independent variables include climate from the four seasons, elevation, surface water flow, farmland size, family size, electricity, four soil types, and the prices of groundnuts, cotton, wheat, cowpea and sorghum.

The regression results in Table 9.5 compare the probability of picking each choice relative to maize. Higher elevation encourages cowpea, sorghum, maize–beans, cowpea–sorghum, maize–groundnut and maize–millet and discourages millet–groundnut. Lower surface water flow increases the chance of selecting maize–beans, cowpea–sorghum, maize–groundnut, maize–millet, millet–groundnut, and fruits–vegetables. Lower flow probably implies that farmers cannot irrigate. Choosing low-water intensive crop combinations is one way for farmers to adapt to rainfed farming in Africa. Farms that have electricity are more likely to choose fruits–vegetables but less likely to choose every other crop. Electricity may help in the production of fruits–vegetables or it may simply signal access to urban markets which often accompanies access to electricity (particularly in the context of rural Africa where electricity distribution networks are predominantly in the vicinity of towns and cities). Farmers whose farms have steep slopes and Ferrasols soils are more likely to pick

Table 9.5 Crop choice model relative to maize in Africa

Variable	Cowpea	Sorghum	Fruits–vegetables	Maize–beans	Cowpea–sorghum	Maize–groundnut	Maize–millet	Millet–groundnut
Temp. winter	-2.81*	1.14*	-1.27*	-0.83	2.12*	-0.87	1.12	-5.41**
	(3.2)	(2.5)	(2.9)	(1.5)	(2.3)	(1.6)	(1.6)	(3.5)
Temp. winter sq.	0.06*	-0.04*	0.04*	0.01	-0.05*	0.00	-0.03	0.13**
	(2.6)	(2.7)	(3.1)	(0.9)	(2.1)	(0.1)	(1.5)	(3.6)
Temp. spring	2.19	-2.73**	-0.10	-0.29	-2.85*	-0.79	-3.09**	6.16**
	(1.8)	(5.2)	(0.2)	(0.4)	(2.6)	(1.2)	(4.3)	(2.9)
Temp. spring sq.	-0.03	0.07**	-0.01	0.01	0.07*	0.048*	0.08**	-0.10*
	(1.0)	(5.4)	(0.7)	(0.6)	(3.0)	(2.7)	(4.6)	(2.4)
Temp. summer	-5.81**	0.89	-0.75	-0.69	-1.06	-3.37**	0.43	6.62**
	(5.0)	(1.5)	(1.1)	(0.7)	(1.0)	(4.4)	(0.6)	(3.9)
Temp. summer sq.	0.11**	-0.01	0.01	0.00	0.03	0.07**	0.00	-0.11**
	(5.3)	(0.9)	(0.4)	(0.1)	(1.4)	(4.3)	(0.0)	(3.7)
Temp. autumn	4.58**	-1.90*	0.39	1.26	-0.15	6.60**	-0.94	-6.31**
	(3.9)	(2.8)	(0.5)	(1.1)	(0.1)	(5.6)	(1.1)	(3.7)
Temp. autumn sq.	-0.10**	0.03*	0.01	-0.01	0.00	-0.16**	0.00	0.11**
	(4.0)	(2.2)	(0.6)	(0.4)	(0.2)	(5.4)	(0.2)	(3.4)
Prec. winter	-0.12**	-0.11**	0.06**	-0.03	-0.17**	-0.02	0.09**	0.03
	(5.7)	(6.5)	(4.1)	(1.7)	(4.8)	(1.7)	(4.1)	(0.7)
Prec. winter sq.	0.001**	0.001**	-0.0002*	0.0002*	0.001**	0.0003**	-0.0005**	0.00
	(7.6)	(7.0)	(2.3)	(2.3)	(8.0)	(4.8)	(3.7)	(1.2)
Prec. spring	0.03	0.06**	-0.06**	0.01	-0.03	0.01	-0.10**	-0.04
	(1.7)	(4.1)	(5.5)	(0.6)	(1.2)	(1.0)	(5.3)	(1.2)
Prec. spring sq.	0.00	-0.0003**	0.0002**	0.00	0.00	0.00	0.0004**	0.00
	(0.9)	(4.7)	(3.3)	(0.5)	(0.0)	(1.3)	(5.3)	(0.6)

Table 9.5 (continued)

Variable	Cowpea	Sorghum	Fruits–vegetables	Maize–beans	Cowpea–sorghum	Maize–groundnut	Maize–millet	Millet–groundnut
Prec. summer	0.21**	-0.05**	0.00	0.01	0.15**	0.02	-0.05**	-0.12**
	(8.2)	(3.9)	(0.3)	(0.6)	(6.7)	(1.6)	(4.2)	(6.9)
Prec. summer sq.	-0.001**	0.0002**	0.00	0.00	-0.001**	0.00	0.0002**	0.0003**
	(7.3)	(4.2)	(0.4)	(0.3)	(5.0)	(1.7)	(3.8)	(4.4)
Prec. autumn	-0.16**	0.01	0.02*	0.00	-0.103**	0.01	0.04*	0.22**
	(7.8)	(0.8)	(2.0)	(0.1)	(4.2)	(1.0)	(3.2)	(7.6)
Prec. autumn sq.	0.001**	-0.0001	-0.00003	0.00	0.0003*	0.00	-0.0001*	-0.0008**
	(7.3)	(1.4)	(0.9)	(0.2)	(2.4)	(0.6)	(3.1)	(7.1)
Mean flow (mm)	0.1*	-0.06*	-0.08**	-0.04	0.00	-0.06	-0.24**	-0.19*
	(2.2)	(2.9)	(3.5)	(1.3)	(0.0)	(1.9)	(3.3)	(2.4)
Elevation (m)	0.0002	0.00	0.0004	0.001*	0.00	0.00	0.001**	0.003**
	(0.7)	(0.3)	(1.7)	(2.6)	(0.1)	(0.7)	(4.4)	(3.8)
Log (farmland)	0.04	0.10	-0.14*	-0.19**	0.19*	0.05	0.187*	0.00
	(0.6)	(1.7)	(2.9)	(3.4)	(2.2)	(1.0)	(2.4)	(0.0)
Log (family size)	0.69**	0.69**	0.41*	0.43*	1.50**	0.81**	0.80**	0.67*
	(3.6)	(5.0)	(3.1)	(3.0)	(7.8)	(7.1)	(4.7)	(3.3)
Electricity dummy	-0.67*	-1.83**	0.19	-0.20	-1.81**	-0.48*	-0.98**	-0.66*
	(2.7)	(7.8)	(1.2)	(1.1)	(6.1)	(3.1)	(4.0)	(2.3)
Steep and Ferrasols	0.99	2.49**	-2.83**	-0.72	0.20	0.75	-0.74	2.06*
	(1.3)	(3.5)	(3.8)	(1.0)	(0.3)	(1.1)	(0.8)	(2.7)
Eutric Gleysols and Solodic Planosols	-1.42**	-0.62*	-1.00*	-1.00*	-1.01*	-0.57*	-0.70*	-1.17*
	(3.7)	(2.4)	(3.1)	(2.6)	(2.9)	(2.6)	(2.2)	(2.5)
Medium texture	-0.76	-2.11*	2.47**	2.06*	-1.90*	-0.75	1.50	-1.87*
	(1.0)	(2.9)	(3.9)	(3.3)	(2.6)	(1.1)	(1.5)	(2.3)
Lithosols								

	(1)	(2)	(3)	(4)	(5)	(6)	(7)	(8)
Chronic Luvisols and Orthic Ferrasols	0.58	−0.39	−0.48	−0.01	−0.28	−0.46	−0.95	−1.95
	(1.0)	(0.9)	(1.2)	(0.0)	(0.2)	(1.8)	(1.9)	(1.6)
Price of groundnut	1.68*	4.37**	1.32*	4.00**	5.66**	1.59**	4.55**	3.54**
	(2.6)	(8.4)	(2.4)	(6.9)	(8.4)	(3.6)	(8.2)	(4.6)
Price of cotton/kg	−5.34**	0.47	−2.55*	−1.92*	0.13	0.50	1.45	−9.78**
	(3.9)	(0.6)	(3.1)	(2.0)	(0.1)	(0.8)	(1.6)	(4.1)
Price of wheat/kg	6.62**	−4.10**	0.33*	0.43	−12.94**	4.33**	0.23	−5.43
	(4.5)	(3.5)	(2.3)	(0.4)	(6.7)	(5.6)	(0.2)	(1.6)
Price of cowpea/kg	−3.26**	0.34	−0.70	−0.32	0.96	−0.42	0.57	−0.61
	(4.6)	(0.7)	(1.3)	(0.6)	(1.4)	(1.1)	(0.9)	(0.5)
Price of sorghum/kg	−0.79	−1.11	1.70*	3.42**	0.92	−0.44	−1.10	−1.05
	(1.0)	(1.5)	(2.6)	(5.2)	(1.2)	(0.7)	(1.4)	(1.0)
Constant	10.20	29.24**	14.33**	1.05	18.84*	−20.70*	22.70**	−53.81**
	(1.4)	(7.5)	(3.3)	(0.2)	(2.7)	(2.8)	(4.3)	(3.3)

Notes: ** significant at 1% level; * significant at 5% level.
Multinomial logistic regression

Number of obs	= 5251
LR chi^2(200)	= 10042
Prob > chi^2	= 0.0000
Pseudo R^2	= 0.45
Log likelihood	= −6184

Source: Kurukulasuriya and Mendelsohn (2008b).

Table 9.6   Marginal effects of climate on choosing each crop relative to maize

Crop	Temperature	Precipitation
Cowpea	1.26	–0.06
Sorghum	–0.08	–0.06
Fruits–vegetables	–0.06	0.01
Maize–beans	–0.08	0.01
Cowpea–sorghum	1.10	–0.17
Maize–groundnut	–0.01	0.02
Maize–millet	0.34	0.0003
Millet–groundnut	0.90	–0.06

Note: Marginal effects estimated from coefficients in Table 9.4.

Source: Kurukulasuriya and Mendelsohn (2008b).

millet–groundnut but less likely to pick cowpea, sorghum, cowpea–sorghum, maize–beans, and fruits–vegetables. Farms with eutric Gleysols and solodic Planosols soils are more likely to have cowpea and maize and less likely to have every other crop. Farms with Lithosols or medium texture soils are more likely to pick cowpea, maize–beans, and fruits–vegetables, but less likely to pick sorghum. Farms with orthic Ferrasols and chromic Luvisols soils are more likely to pick millet–groundnut but less likely to pick sorghum and maize–millet (see Table 9.6).

The choice of different crops is sensitive to seasonal climate variables. Cowpea, millet–maize and sorghum are all sensitive to both temperature and precipitation. In order to get a sense of the impact of climate on crop choice, we calculate the marginal effect of climate on the log odds ratios in Table 9.6. The marginal effect measures how the probability of choosing each crop relative to maize will change as either temperature or precipitation changes. Warmer temperatures reduce the relative probability of maize–millet, maize–groundnut, fruits–vegetables, and especially cowpea, but increase the relative probability of all other crops and especially sorghum. Increased precipitation reduces the relative probability of maize–millet and maize–groundnut, but increases the relative probability of sorghum and cowpea–sorghum.

Another adaptation that is important is whether or not to irrigate a plot (Kurukulasuriya and Mendelsohn, 2008b). A total of 10,900 plots are analyzed in this irrigation analysis. All the plots are devoted to crops. A probit model is used to estimate the choice in Table 9.7. The independent variables include water flow during each season, the square of water flow in

Table 9.7   Probit model of whether to irrigate plots in Africa

Variable	Coefficients	Variable	Coefficients
Temp. winter	0.45*	Gleyic Luvisols: fine,	−7.34*
	(3.00)	undulating	(−1.98)
Temp. winter sq.	0	Eutric Gleysols	−2.54**
	(−0.67)		(−6.57)
Temp. spring	−0.95**	Chromic Cambisols:	−1.54*
	(−6.03)	medium, steep	(−2.51)
Temp. spring sq.	0.01*	Lithosols: coarse, medium,	−5.20*
	(2.48)	fine, steep	(−1.99)
Temp. summer	1.25**	Ferric Luvisols: coarse,	1.09**
	(9.42)	undulating	(8.69)
Temp. summer sq.	−0.02**	Gleyic Luvisols	0.84*
	(−9.43)		(2.60)
Temp. autumn	−0.71**	Gleyic Luvisols: medium,	0.78*
	(−4.25)	undulating	(2.96)
Temp. autumn sq.	0.02**	Chromic Luvisols:	0.53
	(5.54)	medium, undulating, hilly	(1.00)
Precip. winter	−0.01	Luvic Arenosols: coarse,	−4.70**
	(−1.89)	undulating	(−3.76)
Precip. winter sq.	0.00**	Lithosols and Eutric	7.25*
	(5.06)	Gleysols: hilly	(2.54)
Precip. spring	−0.01*	Calcic Yermosols: coarse,	2.74**
	(−2.20)	medium, undulating, hilly	(5.28)
Precip. spring sq.	0.0000025	Eutric Gleysols: coarse,	−2.71
	(0.10)	undulating	(−1.51)
Precip. summer	0.02**	Chromic Vertisols: fine,	0.6
	(6.37)	undulating	(1.06)
Precip. summer sq.	−0.000067**	Chromic Luvisols:	−0.52
	(−5.48)	medium, steep	(−0.31)
Precip. autumn	−0.01**	Dystric Nitosols	−1.02*
Precip. autumn sq.	0.000036**	Lithosols: hilly, steep	−0.06
	(4.00)		(−0.09)
Plot area (ha)	0.000067	Orthic Luvisols: medium,	−1.5
	(0.59)	hilly	(−1.21)
Log (altitude)	0.26**	Flow winter	−1.67
	(8.13)		(−1.73)
Log (household size)	0.09*	Flow winter sq.	−0.8
	(2.03)		(−1.17)
Household with electricity	0.23**	Flow spring	−0.05
(1/0)	(4.33)		(−0.06)
Flow autumn	1.22**	Flow spring sq.	2.12*
	(5.58)		(3.20)

Table 9.7   (continued)

Variable	Coefficients	Variable	Coefficients
Flow autumn sq.	–0.08*	Flow summer	–1.24**
	(–3.28)		(–4.83)
Constant	–4.18**	Flow summer sq.	0.11**
	(–3.73)		(4.55)
N	10915		
Log-likelihood	–2197		
Adjusted R^2	0.55		

Notes:
Dependent variable is whether or not to irrigate a plot.
One asterisk denotes significance at 5% level (*) and two asterisks denotes significance at 1% level (**).
z statistics are in parenthesis.
Soil texture: Coarse soils have less than 18% clay and more than 65% sand, medium soils have less than 35% clay and less than 65% sand, and fine soils have more than 35% clay.
Soil slope: Undulating with less than 8% slope, hilly with slopes between 8% and 30%, and steep with more than 30% slope.

Source: Kurukulasuriya and Mendelsohn (2008a).

each season, altitude, electricity, household size, and 17 soil variables. The seasonal district surface water flow variables and altitude identify choice. Higher altitudes imply rougher terrain, which makes trapping water easier.

The results reveal that higher water flows in every season but winter increase the chance of irrigation. The coefficients on many soils are significant as well. Soils can increase or decrease the probability of irrigation depending on whether they are hilly or undulating (positive) or steep (negative). Often, fine soils are negatively associated with irrigation, and medium soil is positively associated. The effects of types of soils vary depending on slope and soil texture. Electricity is positively associated with irrigation. This may reflect the role of electricity in pumping or just access to markets. Plot size is not related to irrigation choice. Larger households are more likely to irrigate, which suggests that irrigation is labor-intensive on a per hectare basis. Other household variables such as education, age and experience were not significant.

The climate and flow coefficients in Table 9.7 are highly significant. However, with the quadratic functional form, they are hard to interpret directly. Using the coefficients in Table 9.7, we present the mean marginal impact of temperature, precipitation and flow in Table 9.8. The probability of adopting irrigation increases with higher temperatures in each season except in spring. The annual effect of higher temperatures reduces the probability of adopting irrigation. Irrigation allows crops to withstand

Table 9.8   Marginal impact of climate on probability of irrigation in Africa

	Selection model (irrigation choice)		
	Temperature °C	Precipitation mm/mo	Flow million m³/mo
Winter	0.34	–0.002	–2.49
	(0.062)	(0.005)	(0.83)
Spring	–0.52	–0.01	1.47
	(0.068)	(0.004)	(0.85)
Summer	0.08	0.01	–0.9
	(0.58)	(0.002)	(0.28)
Autumn	0.15	–0.002	0.91
	(0.059)	(0.002)	(0.20)
Annual	0.06	–0.01	–1.06
	(0.016)	(0.004)	(0.58)

Note: Standard errors of marginal values are in parenthesis.

Source: Kurukulasuriya and Mendelsohn (2008a).

higher temperatures and the combination of irrigation and higher temperatures allows for multiple seasons. The probability of adopting irrigation falls with more precipitation in every season except summer. With more rain, farmers can grow crops without irrigation, making the cost of irrigation unnecessary. The probability of adopting irrigation falls if there is a uniform annual increase in flow across all seasons. However, this is because flow during the winter season is very harmful, probably causing damage to irrigated systems. Flow during the spring and autumn seasons substantially increases the probability of irrigation. In general, farmers favor irrigation in warmer and drier African climates with good flow in the spring and autumn.

SOUTH AMERICAN STUDIES

We also apply the adaptation choice model to examine a number of different adaptations that farmers make in South America. We rely on a data set explicitly collected to study climate change impacts and adaptations on farmers in South America (Mendelsohn et al., 2007). The data consists of 2500 farms from Argentina, Brazil, Chile, Columbia, Ecuador, Uruguay and Venezuela (Mendelsohn et al., 2007). The countries were chosen to represent the distribution of farms across South America. The farms were selected within each country to represent various climates within each country that can support a farm. Data concerning the land value, farm type choice, species choice and livestock choice were all gathered by farm. Satellite measures of temperature and weather station measures of precipitation were matched with the farm-level data. Soil data from FAO was also included.

The first adaptation decision we examine is the choice of farm type (Mendelsohn and Seo, 2007). Five farm types are considered: a crop-only rainfed farm, a crop-only irrigated farm, a mixed (livestock and crop) rainfed farm, a mixed irrigated farm, and a livestock-only farm. We rely on a multinomial logit regression to estimate what determinants affect these farm choices. The independent variables include: winter and summer climate variables, six soil variables, maize, potato and tomato prices, and a commercial dummy.

The results of the regressions are shown in Table 9.9. Livestock-only farms are the base case. The four regressions represent the likelihood of choosing each of the other farm types relative to livestock-only. The climate coefficients are significant. For the four regressions, the response to higher temperatures is hill-shaped, whereas the response to higher precipitation is U-shaped. Acrisol, Cambisol and Gleysol soils decrease the probability that each farm type is chosen relative to livestock-only farms. Planosol soils, however, increase the probability of crop-only rainfed and mixed rainfed farming being chosen. Commercial farms do not show any substantially different response. Crop-only irrigated farms are more often chosen when tomato prices are high. When maize prices are high, all of the four farm types are chosen more often compared to livestock-only. However, when potato prices are high, livestock-only is chosen more often.

Table 9.10 describes the marginal effects of temperature and precipitation on the choice of farm type. Higher temperatures encourage farmers to move away from crop-only farms towards mixed farms and livestock farms. Higher precipitation pushes farmers to adopt rainfed farming and avoid expensive irrigation investments. The analysis reveals that the choice of farm type is sensitive to climate and will change as climate changes in South America.

Not every farm will make the average changes described in Table 9.10. Adaptations are expected to vary across the landscape. Adaptations depend not only on climate change but also upon the baseline conditions of each farm. For example, farms with different initial climates may respond quite differently to climate change. Increased precipitation may have a minimal effect on the decision to adopt irrigation by a farmer in a wet place but it might change the choice of a farmer in a dry location. Warming at farms that are cool might increase the likelihood of crop-only farming, whereas warming at farms that are already warm might increase the probability of mixed farming.

Table 9.9   Multinomial choice model of farm type in South America

	Crop-only rainfed	Chi-sq	Crop-only irrigated	Chi-sq	Mixed rainfed	Chi-sq	Mixed irrigated	Chi-sq
Intercept	-7.67	10.71	-2.21	1.07	-7.320	12.22	-3.262	2.30
Summer temp.	0.914	14.03	0.568	6.35	0.690	10.78	0.246	1.22
Summer temp. Sq.	-0.0396	30.06	-0.025	16.35	-0.0254	20.99	-0.0144	5.73
Summer prec.	-0.014	7.01	-0.029	28.61	-0.015	8.16	-0.027	22.12
Summer prec. Sq.	0.0000	6.03	0.0001	16.79	0.0000	5.97	0.0001	13.46
Winter temp.	1.027	88.25	0.572	28.23	0.841	75.03	0.993	68.81
Winter temp. Sq.	-0.024	50.19	-0.012	12.64	-0.0191	42.32	-0.0242	42.14
Winter prec.	-0.023	16.29	-0.026	19.76	-0.0191	12.65	-0.0273	20.94
Winter prec. Sq.	0.0001	17.82	0.0001	9.32	0.0001	15.20	0.0001	10.96
Acrisol soils	-0.0218	16.38	-0.0438	20.68	-0.0185	18.30	-0.0349	19.58
Cambisol soils	-0.0267	7.89	-0.0605	21.89	-0.0128	2.21	-0.0277	7.41
Gleysol soils	-0.0239	7.91	-0.0551	14.71	-0.0348	18.06	-0.0446	13.48
Andosol soils	0.0030	0.06	-0.0412	6.50	0.0162	1.81	-0.0124	0.81
Planosol soils	0.0077	4.41	-0.0063	2.24	0.0069	4.40	0.0009	0.05
Xerosol soils	0.0042	0.15	-0.0172	2.29	0.0022	0.06	-0.0213	2.96
Maize price	1.154	28.04	1.029	22.36	31.80	1.352	37.22	
Potato price	-17.269	10.18	-3.463	0.41	5.417	1.41	-0.864	0.00
Tomato price	-3.293	10.06	0.541	0.83	-5.553	32.57	-3.903	10.05
Commercial dummy	0.005	0.00	0.189	3.13	-0.077	0.67	0.176	2.42

Notes:
Values compared to livestock-only choice.
Likelihood ratio test: $P<0.0001$; Lagrange Multiplier test: $P<0.0001$; Wald test: $P<0.0001$.

Source: Mendelsohn and Seo (2007).

Table 9.10   Marginal climate effects on the probability of farm types

	Crop-only rainfed	Crop-only irrigated	Mixed rainfed	Mixed irrigated	Livestock only
Baseline	17.3%	16.5%	43.2%	11.9%	11.1%
Temperature (°C)	−1.5%	−1.7%	+1.2%	+0.5%	+1.5%
Precipitation (mm/mo)	+0.1%	−0.1%	+0.1%	−0.1%	+0.1%

Source: Mendelsohn and Seo (2007).

The second adaptation decision we explore is the choice of livestock species (Seo and Mendelsohn, 2007). We examine the choice of the five most popular species in South America: beef cattle, dairy cattle, sheep, pigs and chickens (Seo and Mendelsohn, 2007). Because many farmers rely on a combination of these animals, we also explore a few popular combinations of these animals. Specifically, we examine beef cattle–dairy cattle, beef cattle–chickens, beef cattle–sheep and pigs–chickens as viable combinations. Together with choosing single species, the total number of choices comes to nine. In Table 9.11, a multinomial logit model is estimated to calculate the probability of choosing each alternative relative to choosing chickens. The independent variables in this regression include winter and summer climate variables, clay soils, electricity, flat terrain and altitude.

The results in Table 9.11 reveal that several of the control variables are significant. Farms with electricity tend to choose beef–sheep more often and beef–dairy cattle less often. When the dominant soil in a district is clay, farmers choose beef, beef–sheep, and beef–dairy more often. When the terrain is flat, they choose dairy more often and they choose beef, beef–chickens, beef–dairy, beef–sheep, and chickens less often. In other words, farmers tend to pick livestock combinations in terrain that is not flat. When the farm is located in high elevations, farmers choose beef cattle, dairy cattle, and chickens more often.

Controlling for these non-climatic factors, Table 9.11 also reveals that not all of the climate coefficients are significant, but there is a significant climate variable explaining every choice. The choices of beef cattle and pigs have several significant climate coefficients. In the case of beef cattle or dairy cattle, the choice function is hill-shaped with regard to summer temperature, which indicates that these animals do not perform well in hot climates. By contrast, the temperature response for sheep is U-shaped, indicating that sheep are preferred in higher temperatures.

Because the shapes are quadratic, the coefficients in Table 9.11 are difficult to interpret directly. Table 9.12 calculates the marginal effects of

Table 9.11   Multinomial logit model of animal species choice in South America

	Beef cattle		Beef cattle – chickens		Beef cattle – dairy cattle	
	Estimate	chi-sq	Estimate	chi-sq	Estimate	chi-sq
Intercept	−10.02	5.40	2.49	0.2	−2.76	0.32
Temp. summer	0.666	2.33	−0.344	0.36	0.030	0.00
Temp. summer sq.	−0.0162	1.61	0.00626	0.14	−0.00392	0.07
Temp. winter	−0.710	8.04	0.067	0.04	−0.246	0.71
Temp. winter sq.	0.0178	5.26	0.00577	0.30	0.014	2.58
Prec. summer	0.0908	10.63	0.0086	0.06	0.0248	0.70
Prec. summer sq.	−0.00022	4.11	−0.0001	0.51	−0.00007	0.39
Prec. winter	0.0747	5.55	0.0286	0.42	0.0387	0.95
Prec. winter sq.	−0.00043	4.70	−0.00044	2.20	−0.00033	1.80
Texture clay	0.468	2.98	0.320	0.67	1.034	6.44
Electricity	0.319	1.01	0.223	0.27	−0.644	2.13
Flat terrain	−0.390	2.63	−0.573	2.80	−0.853	8.19
Altitude	0.00153	5.91	0.000572	0.58	0.000364	0.2

	Beef cattle – sheep		Chickens		Dairy cattle	
	Estimate	chi-sq	Estimate	chi-sq	Estimate	chi-sq
Intercept	−12.87	4.73	0.82	0.04	−7.60	3.07
Temp. summer	0.738	1.65	−0.108	0.06	0.351	0.62
Temp. summer sq.	−0.0135	0.69	0.00168	0.02	−0.0128	0.94
Temp. winter	−0.262	0.71	−0.264	1.02	−0.314	1.48
Temp. winter sq.	−0.0073	0.53	0.0101	1.53	0.00645	0.65
Prec. summer	0.0027	0.01	0.0221	0.94	0.0536	3.71
Prec. summer sq.	0.000154	2.81	−0.00009	0.93	−0.00007	0.43
Prec. winter	0.1239	6.30	0.0364	1.09	0.0843	6.53
Prec. winter sq.	−0.00061	3.85	−0.00025	1.31	−0.00041	4.07
Texture clay	1.352	11.07	0.234	0.65	0.345	1.51
Electricity	0.565	2.67	0.103	0.09	−0.134	0.16
Flat terrain	−0.538	3.56	−0.431	2.8	−0.500	4.05
Altitude	0.000894	0.74	0.00101	2.84	0.00147	6.11

	Pigs		Sheep	
	Estimate	chi-sq	Estimate	chi-sq
Intercept	−11.08	2.65	5.25	0.62
Temp. summer	0.924	1.78	−0.129	0.04
Temp. summer sq.	−0.0274	1.90	0.0131	0.52
Temp. winter	−0.666	3.33	−0.300	0.76

Table 9.11   (continued)

	Pigs		Sheep	
	Estimate	chi-sq	Estimate	chi-sq
Temp. winter sq.	0.0225	4.20	0.0136	1.75
Prec. summer	0.0890	3.29	−0.0649	5.11
Prec. summer sq.	−0.00025	1.97	0.000163	2.55
Prec. winter	0.004	0.01	0.030	0.24
Prec. winter sq.	−0.00006	0.05	−0.00079	3.34
Texture clay	0.469	1.47	−0.152	0.18
Electricity	−0.088	0.03	0.320	0.42
Flat terrain	−0.358	1.13	0.044	0.02
Altitude	0.00131	1.82	−0.00103	0.48

Notes:
Omitted choice is the alternative of both pigs and chickens.
The Likelihood Ratio test: $P<0.0001$; Lagrange Multiplier test: $P<0.0001$; Wald test:
$P<0.0001$.

Source: Seo and Mendelsohn (2007c).

Table 9.12   *Marginal climate effects on animal selection in South America*

	Baseline	Temperature	Precipitation
Beef cattle only	32.71%	3.87%	−0.51%
Beef cattle and chickens	5.58%	−0.64%	0.33%
Beef cattle and dairy cattle	3.59%	−0.35%	0.07%
Beef cattle and sheep	7.55%	0.21%	−0.32%
Chickens	13.55%	−0.12%	0.23%
Dairy cattle	3.09%	0.33%	0.02%
Pigs	15.26%	2.44%	−0.51%
Sheep	12.76%	−5.43%	0.65%
Pigs and chickens	5.91%	−0.31%	0.04%

Note: A unit increase in temperature is 1°C and a unit increase in precipitation is 1mm/mo.

Source: Seo and Mendelsohn (2007c).

temperature and precipitation evaluated at the mean climate of the sample. A small increase in temperature increases the number of farmers adopting beef cattle, dairy cattle and pigs, while it decreases the number of farmers adopting beef–chickens, beef–dairy and sheep. A marginal increase in precipitation decreases the probability of beef cattle, beef–sheep and pigs but it increases beef–chicken, chicken and sheep. These marginal effects are calculated at the mean of the sample and will change in different climates. It is important to interpret these results carefully. The results indicate the direct effect of climate on these animals. However, the results also indicate the indirect effect on the animals through ecosystem changes. A wetter climate for example leads to a higher probability that the ecosystem is a forest and not a savannah or grassland. Wetter and warmer climates may have pest and disease vectors. These long-term indirect effects are reflected in the cross-sectional analysis.

The marginal impact of temperature on livestock choice in Latin America is depicted in Figure 9.3 (see p. 149). The analysis indicates that chickens decline as temperatures warm. Beef cattle at first increase but then decline with higher temperatures. Only sheep really increase with warming. Figure 9.4 (see p. 149) depicts the changes in species with precipitation. Beef cattle fall sharply as precipitation increases, whereas dairy cattle increase sharply.

Livestock adaptations are expected to vary across the landscape. Because the marginal effects depend upon the underlying climate, there will be a different response to climate change in every location. For example, at temperatures below 17°C, sheep probabilities decline with warming but at temperatures above 17°C, they increase. In contrast, beef probabilities increase up to temperatures of 18°C and then they decrease.

The third adaptation decision we examine using the South American data is the choice of crop species (Seo and Mendelsohn, 2008c). We examine the sample of farms that grow crops. The study focuses on the seven major crops grown in South America: fruits–vegetables (31 percent), maize (24 percent), wheat (15 percent), squash (11 percent), rice (8 percent), potato (7 percent), and soybeans (4 percent). The value in parenthesis is the frequency each crop is chosen as the primary (major) crop. Altogether these seven crops generated about 85 percent of the total revenue from crops in the sample.

The multinomial regression shown in Table 9.13 is estimated for the seven crop species. The independent variables in these regressions are winter and summer climate variables, three soils, a computer, age of head, education of head, household size, and maize, wheat, squash, mango and tomato prices. The choice of fruits and vegetables is the base case. Although we explored a four-season model, we were not able to estimate significant results for each season. The South American sample is heavily dominated by tropical observations, where the four seasons are not as distinct as in temperate regions. Other variables such as gender were not significant.

The model in Table 9.13 is significant according to three tests of global significance. Most of the individual coefficients are significant. Positive

Table 9.13 Multinomial logit crop selection for South America

Variable	Maize	Potatoes	Rice	Soybeans	Squash	Wheat
Intercept	4.444	−23.338	−11.823	−6.536	7.774	5.292
Temp. summer	0.025	0.130	4.046*	0.528	−1.255	−0.091
Temp. summer sq.	−0.001	−0.029	−0.151*	−0.010	0.036	0.006
Prec. summer	−0.004	0.234*	0.045*	0.002	0.044*	0.009
Prec. summer sq.	0.00003	−0.002*	−0.00007*	0.00002	−0.00008*	0.00002*
Temp. winter	−0.122	1.844*	−1.380*	0.088	−0.149	−0.561*
Temp. winter sq.	0.003	−0.073*	0.078*	−0.013	−0.002	0.004
Prec. winter	0.005	−0.058*	0.097	0.052**	0.014	−0.005
Prec. winter sq.	0.0001	0.0004*	0.0002	−0.001*	−0.00001	0.0003*
Soil Lithosols	0.013	0.074*	−0.007	0.006	0.011	−0.015
Soil Luvisols	−0.021**	0.039*	0.031	−0.015	−0.152	−0.015
Soil Planosols	0.007	0.024*	−0.052	0.026*	−0.005	0.031*
Computer dummy	0.269	0.761	0.656	−0.108	−0.066	0.057
Age head	0.009	0.038	0.004	0.017	0.005	0.035*
Log household size	−0.909*	−1.215*	−0.937	−0.998*	−0.031	−0.841*
Log education	0.106	1.004*	−0.030	1.085*	0.629*	1.512*
Maize price	0.460*	0.688	1.563*	0.099	−2.45**	0.104
Wheat price	18.724*	−44.745*	27.75**	16.910*	4.026	33.562*
Squash price	−26.401*	79.127	−113.900	−20.114*	−7.197	−27.264*
Mango price	−2.015	−7.649	−11.59**	0.391	−1.387*	−2.009
Tomato price	4.290*	−2.876	10.581	−2.765	0.451	−19.580*

Notes:
Fruit is the omitted choice.
Likelihood ratio test: P<0.0001; Lagrange multiplier test: P<0.0001; Wald test: P<0.0001.
One asterisk denotes significance at 5% level (*) and two asterisks denotes significance at 1% level (**).

Source: Seo and Mendelsohn (2008c).

(negative) coefficients imply that the probability of choosing each crop increases (decreases) as the corresponding variable increases. The coefficient on education is positive and significant for every crop in Table 9.13 except for rice and maize which are not significant. This result implies that lower educated farmers tend to grow fruits–vegetables, the omitted choice. Potatoes are more often chosen when the dominant soil at the farm is a Lithosol. When the dominant soil is a Luvisol, farms tend to choose maize less often, but potatoes more often. Wheat, potatoes and soybeans are more likely to be chosen if a farm has Planosol soils. Farms with computers are more likely to choose potatoes and rice. It is not clear whether this equipment actually enhances the profitability of these crops or whether the computer is a proxy for a missing variable such as technology or market access. Larger families are less likely to choose maize, potatoes, soybeans and wheat. These crops are easily mechanized and so may be selected by farmers with smaller families. Older farmers are more likely to choose wheat. The remaining effects are not significant. Only two of the own prices are significant: maize and wheat; both coefficients are positive as expected, farmers are more likely to choose these crops when their prices are higher. The remaining significant price effects are cross-price terms. When wheat prices are higher, farmers are more likely to pick maize, rice and soybean. When maize prices are higher, they are more likely to pick rice but less likely to pick squash. When squash prices are higher, they are less likely to pick maize, soybean and wheat. Higher tomato prices are associated with higher likelihood to choose maize but lower likelihood to choose wheat. These positive cross-price terms imply a complementarity between the two crops in question.

Maize does not have any significant climate coefficients but all the other crop choices have at least one significant climate coefficient. There are many varieties of maize so that it can effectively grow in many climate zones in South America. The crop is a 'generalist' in the sense that it is grown throughout South America. In contrast, the other crops are more specialists and grow in narrower temperature and precipitation ranges. The climate variables consequently significantly influence crop choice. Rice, for example, is very sensitive to summer and winter temperature and to summer precipitation. Potatoes are very sensitive to winter temperatures and summer and winter precipitation. Squash is sensitive to summer precipitation. Wheat is sensitive to winter temperature and summer and winter precipitation. Fruits–vegetables generally prefer warmer temperatures. Soybeans are sensitive to winter precipitation.

Figure 9.5 (see p. 150) describes the marginal effects of temperature on crop choice in Latin America. The graph reveals that wheat and potatoes are ideal for cool temperatures but not for warmer temperatures. Maize is a generalist, existing across many temperature zones. Rice, squash and especially fruits do much better in hot temperatures. Figure 9.6 (see p. 150) describes the marginal effects of precipitation on crop choice. Rice, squash and fruits are well adapted to wetter conditions. Wheat, soybeans and potatoes prefer dry locations.

In Table 9.14, we calculate the marginal effects of temperature and precipitation on each crop evaluated at the mean climate of the sample. As temperatures warm, farmers tend to choose maize and wheat less often while they choose potatoes, rice, soybean and fruits–vegetables more often. If precipitation increases, farmers move away from maize, wheat and fruits–vegetables to potato, rice and squash. Symmetrically, if precipitation falls, farmers move away from potatoes, rice and squash to maize, wheat and fruits–vegetables.

Table 9.14   Marginal effect of climate change on crop choice in South America

	Maize	Potato	Rice	Soybean	Squash	Wheat	Fruit
Baseline (%)	19.5%	6.8%	4.8%	7.9%	8.0%	14.4%	38.6%
Temperature (°C)	–0.2%	+0.5%	+0.4%	+0.2%	+0.7%	–2.3%	+0.8%
Precipitation (mm/month)	–0.3%	+0.2%	+0.1%	0.0%	+0.1%	–0.1%	–0.2%

Source: Seo and Mendelsohn (2008c).

CONCLUSION

The adaptation analysis reveals that farmers adapt to climate by changing many of the decisions concerning their farms. At the most basic level, they change the type of farm. For example, as it warms, they move from crop-only farms to livestock-only farms. Mixed farms relying on a mixture of crops and livestock are used to diversify farm income. Farmers also adjust their decision over whether or not to use irrigation. Irrigation is clearly one method to overcome the problem of being located in a place that is too dry. However, irrigation also makes farms less vulnerable to warmer temperatures.

Given that a farmer has chosen to grow crops, climate affects the choice of crop species. As temperatures warm, farmers move from crops that like cool temperatures such as wheat, potatoes and sorghum towards crops that can prefer warmer temperatures such as millet, groundnut, and fruits–vegetables. Similarly, as precipitation ranges from dry to wet, farmers will move from millet and cowpea towards fruits, vegetables and rice.

Similarly, for livestock farmers, climate affects the choice of livestock species and herd size. As temperatures warm, farmers in South America move from dairy cattle to beef cattle. As precipitation increases, these same farmers move from beef cattle to dairy cattle. In Africa, as temperatures warm, farmers switch from beef cattle to goats and sheep. As precipitation increases, they move from cattle to chickens and goats. These latter changes are likely to be due to the effect of climate on ecosystems. The dryer climates encourage more grassland, whereas the wetter climates encourage more forest.

REFERENCES

Adams, R., B. McCarl, K. Segerson, C. Rosenzweig, K. Bryant, B. Dixon, R. Conner, R. Evenson and D. Ojima (1999), 'The economic effect of climate change on US agriculture', in R. Mendelsohn and J. Neumann (eds), *The Economic Impact of Climate Change on the United States Economy*, Cambridge, UK: Cambridge University Press, pp. 18–54.

Adams, R., C. Rosenzweig, R.M. Peart, J.T. Ritchie, B.A. McCarl, J.D. Glyer, R.B. Curry, J.W. Jones, K.J. Boote, L.H. Allen (1990), 'Global climate change and US agriculture', *Nature*, **345**, 219–24.

Howitt, R. and E. Pienaar (2006), 'Agricultural impacts', in J. Smith and R. Mendelsohn (eds), *The Impact of Climate Change on Regional Systems: A Comprehensive Analysis of California*, Cheltenham, UK and Northampton, MA, USA: Edward Elgar, pp. 188–207.

Hurd, B., J. Callaway, J. Smith and P. Kirshen (1999), 'Economics effects of climate change on US water resources', in R. Mendelsohn and J. Neumann (eds), *The Impact of Climate Change on the United States Economy*, Cambridge, UK: Cambridge University Press, pp. 133–77.

Kaiser, H.M., S.J. Riha, D.S. Wilkes, D.G. Rossiter and R.K. Sampath (1993), 'A farm-level analysis of economic and agronomic impacts of gradual warming', *American Journal of Agricultural Economics*, **75**, 387–98.

Kelly, D.L., C.D. Kolstad and G.T. Mitchell (2005), 'Adjustment costs from environmental change', *Journal of Environmental Economics and Management*, **50**(3), 468–95.

Kurukulasuriya, P. and R. Mendelsohn (2007), 'A Ricardian analysis of the impact of climate change on African cropland', World Bank Policy Research Working Paper 4305, Washington, DC, USA: World Bank.

Kurukulasuriya, P. and R. Mendelsohn (2008a), 'Modeling endogenous irrigation: the impact of climate change on farmers in Africa', World Bank Policy Research Working Paper 4278, Washington, DC, USA: World Bank.

Kurukulasuriya, P. and R. Mendelsohn (2008b), 'Crop switching as an adaptation strategy to climate change', *African Journal of Agriculture and Resource Economics*, **2**, 105–26.

Lund, J., T. Zhu, S. Tanaka and M. Jenkins (2006), 'Water resource impacts', in J. Smith and R. Mendelsohn (eds), *The Impact of Climate Change on Regional Systems: A Comprehensive Analysis of California*, Cheltenham, UK and Northampton, MA, USA: Edward Elgar, pp. 165–87.

Mendelsohn, R. and S.N. Seo (2007), 'An integrated farm model of crops and livestock: modeling Latin American agricultural impacts and adaptation to climate change', World Bank Policy Research Working Paper 4161, Washington, DC, USA: World Bank.

Mendelsohn, R., A.F. Avila and S.N. Seo (2007), 'Synthesis of the Latin American Results', Montevideo, Uruguay: PROCISUR.

Seo, S.N. and R. Mendelsohn (2008a), 'Measuring impacts and adaptation to climate change: a structural Ricardian model of African livestock management', *Agricultural Economics*, **38**, 150–65.

Seo, S.N. and R. Mendelsohn (2008b), 'Climate change impacts and adaptations on animal husbandry in Africa', *African Journal of Agriculture and Resource Economics*, **2**, 65–82.

Seo, S.N. and R. Mendelsohn (2008c), 'An analysis of crop choice: adapting to climate change in Latin American farms', *Ecological Economics*, **67**, 109–16.

Seo, S.N. and R. Mendelsohn (2008d), 'Can farmers adapt to climate change?: Livestock choice and herd size in Africa', Yale FES Working Paper, New Haven, CT, USA.

Seo, S.N. and R. Mendelsohn (2007a), 'Climate change impacts on animal husbandry in Africa: a Ricardian analysis', World Bank Policy Research Working Paper 4261, Washington, DC, USA: World Bank.

Seo, S.N. and R. Mendelsohn (2007b), 'A Ricardian analysis of the impact of climate change on Latin American farms', World Bank Policy Research Working Paper 4163, Washington, DC, USA: World Bank.

Seo, S.N. and R. Mendelsohn (2007c), 'An analysis of livestock choice: adapting to climate change in Latin American farms', World Bank Policy Research Working Paper 4164, Washington, DC, USA: World Bank.

Seo, S.N. and R. Mendelsohn (2007d), 'An analysis of crop choice: adapting to climate change in Latin American farms', World Bank Policy Research Working Paper 4162, Washington, DC, USA: World Bank.

Strzepek, K., D. Yates and D. El Quosy (1996), 'Vulnerability assessment of water resources in Egypt to climatic change in the Nile Basin', *Climate Research*, **6**: 89–95.

10.   Structural Ricardian studies

The structural Ricardian modeling approach discussed in Chapter 6 has been applied to choices in both South America and Africa. In Africa, the technique has been used to examine the choice of irrigation, crops and livestock. In South America, the method has been used to examine the choice of farm type (whether to irrigate and whether to grow crops and/ or livestock).

The African data is based on a sample of over 9000 farmers collected in 11 countries in Africa: Burkina Faso, Cameroon, Egypt, Ethiopia, Ghana, Kenya, Niger, Senegal, South Africa, Zambia and Zimbabwe (Dinar et al., 2008). Data includes information on net revenues from crops and livestock as well as choices concerning crop species, irrigation and livestock species. Data also includes substantial information about farmers including education, age, household size and gender. Temperature data from satellites (Weng and Grody, 1998) and precipitation data from weather stations was matched to each farm (World Bank, 2003). Soil data from the FAO was then matched to each farm (FAO, 2003). Finally, water flow data from hydrological modeling was added (IWMI, 2003).

The South America data is based on a sample of over 2000 farmers collected across seven countries in South America: Argentina, Brazil, Chile, Columbia, Ecuador, Uruguay and Venezuela (Mendelsohn et al., 2007). As with the African study, the South American study includes extensive economic data as well as climate and soil information. Data was not available concerning the hydrology. The South American study, however, has information about land values and more complete information about the land used for livestock.

For all the structural Ricardian models, we first discuss the estimation of the models, then the marginal climate impacts, and finally the predicted impacts of two future climate scenarios. The two scenarios are predictions for 2100 from two climate models: the Parallel Climate Model (PCM) (Washington et al., 2000) and the Canadian Climate Centre model (CCC) (Boer et al., 2000). The two climate scenarios were selected to reflect the range of likely impacts predicted by climate experts (IPCC, 2007). In Africa, PCM is a mild and wet scenario with a predicted average increase of 2.3°C, and CCC is a very hot and dry scenario with a predicted temperature increase of 6.5°C. In South America, the average temperature increase with PCM is 2.0°C, and with CCC, it is 5.1°C.

IRRIGATION CHOICE

This irrigation study of Africa estimates a model of the choice of whether farmers adopt irrigation and then a model of their conditional income (Kurukulasuriya and Mendelsohn, 2008a). The irrigation model begins with a probit selection equation that estimates the probability of a farmer selecting irrigation or not (rainfed). Altitude and the amount of water flow in the district are used to identify the choice equation. This equation is followed by a conditional income equation for irrigated and rainfed land. This second equation estimates the net revenue per hectare that each plot of land would earn, conditional on whether or not it is irrigated. The farmer may be aware of factors unknown to the modeler and so it is likely that the error terms of the two regressions will be correlated (Heckman (1979) selection bias). A selection error term is introduced into the conditional income equation to correct for this bias.

Table 10.1 presents the probit model of whether or not a plot is irrigated. There are 10,915 observed plots in the regression. The explanatory variables in the first stage include seasonal climate variables, farm characteristics, soils and seasonal water flow. Most of the control variables are significant. Higher altitudes in Africa generally imply more fertile land, making irrigation more attractive. Higher water flows in each season except winter make irrigation more attractive. Water flows are not measures of the water used by each farmer but rather the exogenous water flow in a district. Soils can increase or decrease the probability of irrigation depending on whether they are hilly or undulating (positive) or steep (negative). Often, fine soils are negatively associated with irrigation and medium soils are positively associated. The effects of types of soils vary depending on slope and soil texture. Electricity is positively associated with irrigation. This may reflect the role of electricity in pumping or just access to markets. Plot size is not related to irrigation choice. Larger households are more likely to irrigate, which suggests that irrigation is labor-intensive on a per hectare basis. Other household variables such as education, age and experience were not significant.

The climate coefficients are highly significant. The quadratic functional form, however, makes them hard to interpret. We present the mean marginal impact of temperature, precipitation, and flow on irrigation choice below in Table 10.2. The probability of adopting irrigation increases with higher temperatures in each season except in spring. The

Table 10.1   Probit model choice of irrigation in Africa

Variables	Coefficient	Variables	Coefficient
Winter temp.	0.45*	Winter prec.	−0.01
	(3.00)		(−1.89)
Winter temp. sq.	0.00	Winter prec. sq.	0.00**
	(−0.67)		(5.06)
Spring temp.	−0.95**	Spring prec.	−0.01*
	(−6.03)		(−2.20)
Spring temp. sq.	0.01*	Spring prec. sq.	0.0000025
	(2.48)		(0.10)
Summer temp.	1.25**	Summer prec.	0.02**
	(9.42)		(6.37)
Summer temp. sq.	−0.02**	Summer prec. sq.	−0.000067**
	(−9.43)		(−5.48)
Autumn temp.	−0.71**	Autumn prec.	−0.01**
	(−4.25)		(−3.25)
Autumn temp. sq.	0.02**	Autumn prec. sq.	0.000036**
	(5.54)		(4.00)
Plot area (ha)	0.000067	Electricity (1/0)	0.23**
	(0.59)		(4.33)
Log (elevation)	0.26**	Gleyic Luvisols FU	−7.34*
	(8.13)		(−1.98)
Log (household size)	0.09*	Eutric Gleysols	−2.54**
	(2.03)		(−6.57)
Chromic Cambisols: medium, steep	−1.54*	Gleyic Luvisols	0.84*
	(−2.51)		(2.60)
Lithosols	−5.20*	Gleyic Luvisols MU	0.78*
	(−1.99)		(2.96)
Ferric Luvisols	1.09**	Chromic Luvisols	0.53
	(8.69)		(1.00)
Luvic Arenosols	−4.70**	Eutric Gleysols CU	−2.71
	(−3.76)		(−1.51)
Lithosols and Eutric Gleysols: hilly	7.25*	Chromic Vertisols	0.6
	(2.54)		(1.06)
Calcic Yermosols: coarse, medium, undulating, hilly	2.74**	Chromic Luvisols MU	−0.27
	(5.28)		(−0.61)
Chromic Luvisols: medium, steep	−0.52	Lithosols: hilly, steep	−0.06
	(−0.31)		(−0.09)
Dystric Nitosols	−1.02*	Orthic Luvisols MH	−1.5
	(−2.06)		(−1.21)
Winter flow	−1.67	Summer flow	−1.24**
	(−1.73)		(−4.83)

Table 10.1 (continued)

Variables	Coefficient	Variables	Coefficient
Winter flow sq.	−0.8	Summer flow sq.	0.11**
	(−1.17)		(4.55)
Spring flow	−0.05	Autumn flow	1.22**
	(−0.06)		(5.58)
Spring flow sq.	2.12*	Autumn flow sq.	−0.08*
	(3.20)		(−3.28)
Constant	−4.18**		
	(−3.73)		
Log-likelihood	−2197	R^2	0.55

Notes:
There are 10 915 observations. t-statistics are in parentheses.
FU = fine undulating soil; MU = medium undulating soil; CU = clay undulating soil.
* indicates significance at 5% level; ** indicates significance at 1% level.

Source: Kurukulasuriya and Mendelsohn (2008a).

Table 10.2 Marginal irrigation selection results for Africa

Seasons	Marginal selection (Irrigation=1)		
	Temperature °C	Precipitation mm/mo	Flow million m³/mo
Winter	0.34	−0.002	−2.49
	(0.062)	(0.005)	(0.83)
Spring	−0.52	−0.01	1.47
	(0.068)	(0.004)	(0.85)
Summer	0.08	0.01	−0.9
	(0.58)	(0.002)	(0.28)
Autumn	0.15	−0.002	0.91
	(0.059)	(0.002)	(0.20)
Annual	0.06	−0.01	−1.06
	(0.016)	(0.004)	(0.58)

Note: Calculates increased probability of irrigation per unit of climate at African mean from coefficients in Table 10.1. t-statistics are in parentheses.

Source: Kurukulasuriya and Mendelsohn (2008a).

annual effect of higher temperatures reduces the probability of adopting irrigation. Although irrigated crops may be able to withstand higher temperatures, warmer climates require more water for irrigation, making them less attractive. The probability of adopting irrigation falls with more precipitation in every season except summer. With more rain, farmers can grow crops without irrigation, making the cost of irrigation unnecessary. The probability of adopting irrigation falls if there is a uniform annual increase in flow across all seasons. However, this is because flow during the winter season is very harmful, probably causing damage to irrigated systems. Flow during the spring and autumn seasons substantially increases the probability of irrigation. In general, farmers favor irrigation in warmer and drier African climates with good flow in the spring and autumn.

The second stage model of net revenue conditional on irrigation choice is shown in Table 10.3. The dependent variable is annual net revenue per hectare and the independent variables include climate, soils and other control variables (but not altitude and flow). The regressions correct for selection bias using the inverse Mills ratio. However, the coefficient on the estimated Mills ratio is not significant. Comparing the corrected model with an OLS model reveals that they are not significantly different. There is little evidence of selection bias in this African irrigation study.

Farm size is significant and negative for both the irrigated and rainfed plots in Table 10.3. Larger plots have lower net revenue per hectare. This is likely to be due to the omission of household labor as a cost in net revenue (a measurement bias). Household labor per hectare will tend to be greater in smaller plots. The result may also be due to higher management intensity on smaller plots (a real effect). We also include a dummy variable that denotes whether or not a farm has electricity. Electrified farms have higher net revenues than farms that do not have electricity in both the irrigated and rainfed models. Electrification might directly enhance productivity and earnings or it may simply be a proxy for farms that are closer to markets or more modern. Farms with larger households have higher net revenue in both samples, but the coefficient is significant in only the irrigated sample.

The second stage regressions also give important insights into the climate sensitivity of farms. The results show that net revenues of rainfed and irrigated farms are both sensitive to climate but have different climate responses. In order to interpret the climate coefficients, the mean marginal impact on net revenue is presented in Table 10.4. Annual warmer temperatures do not have a significant effect on irrigated farm net revenue as seasonal effects are offsetting. Annual precipitation also does not have a significant effect on irrigated farm income, although wetter summers are beneficial and wetter autumn are harmful. Warmer annual temperatures reduce the income from rainfed plots with harmful effects from warmer springs and autumn but offsetting beneficial effects from warmer winters and summers. Although these seasonal results are quite different from

Table 10.3   Conditional income regressions for Africa

Variables	Irrigated	Rainfed	Variables	Irrigated	Rainfed
Winter temp.	97.0	–128.2*	Winter prec.	12.03	–2.60*
	(0.60)	(–2.51)		(1.80)	(–2.20)
Winter temp. sq.	–1.69	4.33**	Winter prec. sq.	–0.06	0.02*
	(–0.44)	(3.31)		(–1.44)	(2.75)
Spring temp.	–93.2	4.3	Spring prec.	–10.31	3.71**
	(–0.47)	(0.05)		(–1.61)	(3.41)
Spring temp. sq.	–0.60	–1.98	Spring prec. sq.	0.09*	–0.01
	(–0.15)	(–1.09)		(2.30)	(–1.44)
Summer temp.	1188.1**	214.2*	Summer prec.	26.25**	4.09**
	(3.42)	(3.24)		(4.98)	(6.08)
Summer temp. sq.	–18.16*	–2.99*	Summer prec. sq.	–0.10**	–0.02**
	(–3.02)	(–2.36)		(–4.88)	(–5.29)
Autumn temp.	–1580.4**	–82.6	Autumn prec.	–25.35**	–1.21*
	(–3.63)	(–1.47)		(–5.00)	(–2.15)
Autumn temp. sq.	29.43**	1.13	Autumn prec. sq.	0.08**	0.01**
	(3.47)	(0.95)		(4.98)	(5.53)
Plot area (ha)	–0.15*	–0.29**	Chromic	–1910.2*	–708.3**
	(–2.39)	(–4.54)	Vertisols: fine, undulating	(–2.81)	(–3.51)
Log (household size)	41.68	22.46*	Chromic Luvisols: medium fine, undulating		–315.1** (–8.79)
	(.74)	(2.05)			
With electricity (1/0)	387.4**	124.1**	Chromic Luvisols: medium, steep	–6510.5* (–2.94)	
	(3.66)	(7.84)			
Gleyic Luvisols		–111.2*	Dystric Nitosols	7528.3**	
		(–2.89)		(5.39)	
Luvic Arenosols: coarse, undulating		–357.0** (–4.60)	Lithosols: hilly, steep	–877.9* (–3.29)	–352.7** (–8.40)
Eutric Gleysols: coarse, undulating	–1554.0* (–2.25)	–405.4** (–4.54)	Orthic Luvisols: medium, hilly		–1885.7** (–3.81)
Constant	4361.7*	–295.6	Inverse Mills ratio	–102.2	–7.8
	(2.72)	(–0.66)		(–0.80)	(–1.35)
Adjusted R²	0.25	0.16	F-test	68.47	49.41
N	1787	9128			

Notes:
Dependent variable is net revenue per hectare.
* indicates significance at 5% level; ** indicates significance at 1% level.

Source: Kurukulasuriya and Mendelsohn (2008a).

Table 10.4   Marginal conditional income results in Africa

Seasons	Marginal income		Marginal income	
	Irrigated farms		Rainfed farms	
	Temperature °C	Precipitation mm/mo	Temperature °C	Precipitation mm/mo
Winter	45	8	55	–2
	(128)	(9)	(14)	(1)
Spring	–108	–6.0	–97	3
	(140)	(6)	(16)	(1)
Summer	314	17	68	1
	(130)	(5)	(14)	(0.3)
Autumn	–249	–18	–33	1
	(130)	(5)	(15)	(0.4)
Annual	1	1	–7	3
	(25)	(10)	(4)	(0.6)

Note: The marginal income effects are calculated from coefficients in Table 10.3. The standard deviation (in parentheses) was calculated using bootstrapping (350 repetitions).

Source: Kurukulasuriya and Mendelsohn (2008a).

temperate climate findings (Mendelsohn et al., 1994; Mendelsohn and Dinar, 2003), one must remember that spring is often the hottest season in Africa. More annual precipitation increases rainfed plot income. Precipitation is especially beneficial in the spring and harmful only in the winter. The results indicate that rainfed and irrigated plots have very different climate sensitivities.

We now examine the projected impacts given two possible future climate scenarios, in Table 10.5. According to the coefficients in Table 10.1, the probability of selecting irrigation increases dramatically in the PCM climate scenario. Although the seasonal temperature effects are largely off-setting, the large increase in winter precipitation encourages many farms to switch to irrigation. If flow remained the same in the new scenario, the probability of irrigation in the sample would rise dramatically to 56 percent. When the predicted change in water flow is taken into account, the increase in irrigation in the PCM scenario is 44 percent. Note that new water storage facilities, which could hold back winter flows and make them available in the spring and summer, might convert harmful winter flows into beneficial spring and summer flows. We do not take changes in water management techniques into account but they appear to be poten-tially very promising in sub-Saharan Africa. With the CCC scenario, the

Table 10.5   African welfare results with two climate change scenarios

Irrigation	Exogenous	Endogenous (with flow constant)	Endogenous (with flow adjusting to climate)
PCM Scenario			
Probability of irrigation	16%	56%	44%
Δ in expected welfare ($/ha)	44*	169*	115.5*
	(119)	(314)	(299)
Δ in expected welfare (%)	+9%	+35%	+24%
CCC Scenario			
Probability of irrigation	16%	14%	14%
Δ in expected welfare ($/ha)	–276*	–278*	–288*
	(75)	(75)	(68)
Δ in expected welfare (%)	–57%	–58%	–60%

Notes:
Baseline irrigation probability is 16% and baseline net revenue is $483/ha.
* indicates significance at 5% level.
Standard errors are in parentheses.

Source: Kurukulasuriya and Mendelsohn (2008a).

seasonal temperature, precipitation and flow effects are offsetting and the probability of irrigation falls slightly.

Multiplying through the probability of selecting irrigation times the conditional income for each equation provides an estimate of welfare (expected net revenue per hectare). We also calculate the change in welfare per hectare for each climate scenario in Table 10.5. The first column presents the change in welfare effects assuming that rainfed and irrigated farms stay as they are now. That is, the probability of irrigation does not change. The remaining two columns allow the probability of irrigation to adjust to the climate scenario. The second column assumes that water flows do not change and the third column adjusts to likely changes in water flow. The exogenous estimates in the first column grossly underestimate the benefits of the PCM scenario because they do not take into account the large increase in irrigation induced by this wetter scenario. The exogenous model consequently predicts that PCM would lead to only a small benefit of 9 percent whereas the endogenous model predicts a benefit of 35 percent. Adjusting for the harmful effect of winter flows, the endogenous model still predicts that the PCM scenario would lead to a welfare benefit of 24 percent. With the hotter and dryer climate scenarios, there is not as big a difference across the results because irrigation probabilities do not change much. Although the PCM scenario actually predicts substantial gains for African farmers, the much hotter and dryer climate scenario predicts large losses. Without additional water, irrigation will not help farmers escape the very high temperatures of the CCC climate scenario.

CROP CHOICE

This study analyzes the eight most popular crop choices by African farmers: maize (1071 observations), maize–groundnut (811), cowpea–sorghum (666), sorghum (569), millet–groundnut (568), fruit–vegetables (556), maize–beans (399), cowpea (388), and maize–millet (331) (Kurukulasuriya and Mendelsohn, 2008b). Approximately three-quarters of all farmers in the sample chose one of these combinations. Table 10.6 presents the multinomial logit (MNL) regression results of crop choice on climate, flow, soils and other control variables. Higher elevation encourages cowpea, sorghum, maize–beans, cowpea–sorghum, maize–groundnut and maize–millet and discourages millet–groundnut. Lower flow is associated with farmers choosing maize–beans, cowpea–sorghum, maize–groundnut, maize–millet, millet–groundnut and fruits–vegetables. Lower flow probably indicates that it will be more expensive for farmers to irrigate. Choosing low water-intensive crop combinations is one way for farmers to adapt to rainfed farming. Farms that have electricity are more likely to choose fruit–vegetables but less likely to choose every other crop. Electricity may help in the production of fruit–vegetables or it may simply signal access to urban markets. Farmers whose farms have steep slopes and fine-texture soils are more likely to pick millet–groundnut but less likely to pick cowpea, sorghum, cowpea–sorghum, maize–beans and fruits–vegetables. Farms with eutric Gleysol and solodic Planosol soils are more likely to have cowpea and maize and less likely to have every other crop. Farms with Lithosol or medium texture soils in steep areas are more likely to pick cowpea, maize–beans and fruit–vegetables, but less likely to pick sorghum. Finally, those farms with orthic Ferrasol and chromic Luvisol soils are more likely to pick millet–groundnut but less likely to pick sorghum and maize–millet.

From the perspective of this study, the most important coefficients in Table 10.6 concern the seasonal climate. The choice of different crops is sensitive to seasonal climate variables. For example, in comparison to maize, cowpea reacts to summer and autumn temperatures and winter, summer and autumn precipitation, whereas sorghum reacts to winter, spring and autumn temperatures and precipitation in every season. Millet–maize in comparison reacts to winter and spring temperature and precipitation in

Table 10.6 Multinomial logit crop choice model in Africa

	Cowpea	Sorghum	Fruits–vegetables	Maize–beans	Cowpea–sorghum	Maize–groundnut	Maize–millet	Millet–groundnut
Winter temp.	-2.81*	1.14*	-1.27*	-0.83	2.12*	-0.87	1.12	-5.41**
	(3.2)	(2.5)	(2.9)	(1.5)	(2.3)	(1.6)	(1.6)	(3.5)
Winter temp. sq.	0.06*	-0.04*	0.04*	0.01	-0.05*	0.00	-0.03	0.13**
	(2.6)	(2.7)	(3.1)	(0.9)	(2.1)	(0.1)	(1.5)	(3.6)
Spring temp.	2.19	-2.73**	-0.10	-0.29	-2.85*	-0.79	-3.09**	6.16**
	(1.8)	(5.2)	(0.2)	(0.4)	(2.6)	(1.2)	(4.3)	(2.9)
Spring temp. sq.	-0.03	0.07**	-0.01	0.01	0.07*	0.048*	0.08**	-0.10*
	(1.0)	(5.4)	(0.7)	(0.6)	(3.0)	(2.7)	(4.6)	(2.4)
Summer temp.	-5.81**	0.89	-0.75	-0.69	-1.06	-3.37**	0.43	6.62**
	(5.0)	(1.5)	(1.1)	(0.7)	(1.0)	(4.4)	(0.6)	(3.9)
Summer temp. sq.	0.11**	-0.01	0.01	0.00	0.03	0.07**	0.00	-0.11**
	(5.3)	(0.9)	(0.4)	(0.1)	(1.4)	(4.3)	(0.0)	(3.7)
Autumn temp.	4.58**	-1.90*	0.39	1.26	-0.15	6.60**	-0.94	-6.31**
	(3.9)	(2.8)	(0.5)	(1.1)	(0.1)	(5.6)	(1.1)	(3.7)
Autumn temp. sq.	-0.10**	0.03*	0.01	-0.01	0.00	-0.16**	0.00	0.11**
	(4.0)	(2.2)	(0.6)	(0.4)	(0.2)	(5.4)	(0.2)	(3.4)
Winter prec.	-0.12**	-0.11**	0.06**	-0.03	-0.17**	-0.02	0.09**	0.03
	(5.7)	(6.5)	(4.1)	(1.7)	(4.8)	(1.7)	(4.1)	(0.7)
Winter prec. sq.	0.001**	0.001**	-0.0002*	0.0002*	0.001**	0.0003**	-0.0005**	0.00
	(7.6)	(7.0)	(2.3)	(2.3)	(8.0)	(4.8)	(3.7)	(1.2)
Spring prec.	0.03	0.06**	-0.06**	0.01	-0.03	0.01	-0.10**	-0.04
	(1.7)	(4.1)	(5.5)	(0.6)	(1.2)	(1.0)	(5.3)	(1.2)
Spring prec. sq.	0.00	-0.0003**	0.0002**	0.00	0.00	0.00	0.0004**	0.00
	(0.9)	(4.7)	(3.3)	(0.5)	(0.0)	(1.3)	(5.3)	(0.6)

	(1)	(2)	(3)	(4)	(5)	(6)	(7)	(8)
Summer prec.	0.21** (8.2)	−0.05** (3.9)	0.00 (0.3)	0.01 (0.6)	0.15** (6.7)	0.02 (1.6)	−0.05** (4.2)	−0.12** (6.9)
Summer prec. sq.	−0.001** (7.3)	0.0002** (4.2)	0.00 (0.4)	0.00 (0.3)	−0.001** (5.0)	0.00 (1.7)	0.0002** (3.8)	0.0003** (4.4)
Autumn prec.	−0.16** (7.8)	0.01 (0.8)	0.02* (2.0)	0.00 (0.1)	−0.103** (4.2)	0.01 (1.0)	0.04* (3.2)	0.22** (7.6)
Autumn prec. sq.	0.001** (7.3)	−0.0001 (1.4)	−0.00003 (0.9)	0.00 (0.2)	0.0003* (2.4)	0.00 (0.6)	−0.0001* (3.1)	−0.0008** (7.1)
Mean flow (mm)	0.1* (2.2)	−0.06* (2.9)	−0.08** (3.5)	−0.04 (1.3)	0.00 (0.0)	−0.06 (1.9)	−0.24** (3.3)	−0.19* (2.4)
Elevation (m)	0.0002 (0.7)	0.00 (0.3)	0.0004 (1.7)	0.001* (2.6)	0.00 (0.0)	0.00 (0.7)	0.001** (4.4)	0.003** (3.8)
Log (farmland area)	0.04 (0.6)	0.10 (1.7)	−0.14* (2.9)	−0.19** (3.4)	0.19* (2.2)	0.05 (1.0)	0.187* (2.4)	0.00 (0.0)
Log (household size)	0.69** (3.6)	0.69** (5.0)	0.41* (3.1)	0.43* (3.0)	1.50** (7.8)	0.81** (7.1)	0.80** (4.7)	0.67** (3.3)
Electricity dummy	−0.67* (2.7)	−1.83** (7.8)	0.19 (1.2)	−0.20 (1.1)	−1.81** (6.1)	−0.48* (3.1)	−0.98** (4.0)	−0.66* (2.3)
Steep and Ferrasols	0.99 (1.3)	2.49** (3.5)	−2.83** (3.8)	−0.72 (1.0)	0.20 (0.3)	0.75 (1.1)	−0.74 (0.8)	2.06* (2.7)
Eutric Gleysols and solodic Phanosols	−1.42** (3.7)	−0.62* (2.4)	−1.00* (3.1)	−1.00* (2.6)	−1.01* (2.9)	−0.57* (2.6)	−0.70* (2.2)	−1.17* (2.5)
Medium texture Lithosols	−0.76 (1.0)	−2.11** (2.9)	2.47** (3.9)	2.06* (3.3)	−1.90* (2.6)	−0.75 (1.1)	1.50 (1.5)	−1.87* (2.3)
Chromic Luvisols and orthic Ferrasols	0.58 (1.0)	−0.39 (0.9)	−0.48 (1.2)	−0.01 (0.0)	−0.28 (0.2)	−0.46 (1.8)	−0.95 (1.9)	−1.95 (1.6)

Table 10.6 (continued)

	Cowpea	Sorghum	Fruits–vegetables	Maize–beans	Cowpea–sorghum	Maize–groundnut	Maize–millet	Millet–groundnut
Price of groundnut	1.68*	4.37**	1.32*	4.00**	5.66**	1.59**	4.55**	3.54**
	(2.6)	(8.4)	(2.4)	(6.9)	(8.4)	(3.6)	(8.2)	(4.6)
Price of cotton/kg	−5.34**	0.47	−2.55*	−1.92*	0.13	0.50	1.45	−9.78**
	(3.9)	(0.6)	(3.1)	(2.0)	(0.1)	(0.8)	(1.6)	(4.1)
Price of wheat/kg	6.62**	−4.10**	0.33*	0.43	−12.94**	4.33**	0.23	−5.43
	(4.5)	(3.5)	(2.3)	(0.4)	(6.7)	(5.6)	(0.2)	(1.6)
Price of cowpea/kg	−3.26**	0.34	−0.70	−0.32	0.96	−0.42	0.57	−0.61
	(4.6)	(0.7)	(1.3)	(0.6)	(1.4)	(1.1)	(0.9)	(0.5)
Price of sorghum/kg	−0.79	−1.11	1.70*	3.42**	0.92	−0.44	−1.10	−1.05
	(1.0)	(1.5)	(2.6)	(5.2)	(1.2)	(0.7)	(1.4)	(1.0)
Constant	10.20	29.24**	14.33**	1.05	18.84*	−20.70*	22.70**	−53.81**
	(1.4)	(7.5)	(3.3)	(0.2)	(2.7)	(2.8)	(4.3)	(3.3)

Notes:
Base category crop: Maize.
** indicates significance at 1% level; * indicates significance at 5% level.
Multinomial logistic regression

Number of obs	= 5251
LR chi^2(200)	= 10042
Prob > chi^2	= 0.0000
Pseudo R^2	= 0.45
Log likelihood	= −6184

Source: Kurukulasuriya and Mendelsohn (2008b).

Table 10.7   Marginal effects of climate on crop selection in Africa

Crop	Temperature	Precipitation
Cowpea	1.26	–0.06
Maize		
Sorghum	–0.08	–0.06
Fruits–Vegetables	–0.06	0.01
Maize–Beans	–0.08	0.01
Cowpea–Sorghum	1.10	–0.17
Maize–Groundnut	–0.01	0.02
Maize–Millet	0.34	0.0003
Millet–Groundnut	0.90	–0.06

Note: Marginal effects on the log odds of selection estimated from coefficients in Table 10.6.

Source: Kurukulasuriya and Mendelsohn (2008b).

all seasons. In order to get a sense of the impact of climate on crop choice, we calculate the marginal climate impacts in Table 10.7. The marginal climate impacts measures the change in the log odds of selecting a crop relative to maize for each increment of either annual temperature or precipitation. Warmer temperatures reduce the selection of maize–millet, maize–groundnut, fruit–vegetables, and especially cowpea but increase the selection of all other crops and especially sorghum. Increased precipitation reduces the probability of maize, maize–millet and maize–groundnut, but increases the probability of sorghum and cowpea–sorghum.

Table 10.8 presents the results of the second stage conditional income regressions by crop. Selection bias is controlled using Durbin–McFadden selection bias correction terms. Many of the seasonal climate variables in Table 10.8 are statistically significant, although many of the squared terms are not significant. There are fewer observations in each conditional income equation compared to the choice model which probably explains why the individual coefficients of the second stage model are less significant. The significant selection term coefficients in Table 10.8 are all negative. A negative selection term coefficient implies there is a negative correlation between the errors in the selection and conditional income equations. For example, the selection term in the cowpea regression for maize–millet is negative. This implies that if a farmer chooses to grow cowpea, but the selection model predicted he/she would choose maize–millet, the farmer will earn more than expected growing cowpea. The fact that the selection terms are significant suggests that there are issues with sample selection in the African crop choice study.

Table 10.9 presents the marginal impact on conditional income of

Table 10.8 Conditional income by crop in Africa

Independent variable	Cowpea Coef.	Maize Coef.	Sorghum Coef.	Fruits–vegetables Coef.	Maize–beans Coef.	Cowpea–sorghum Coef.	Maize–groundnut Coef.	Maize–millet Coef.	Millet–groundnut Coef.
Price of cowpea/kg	296.2* (–2.12)					40.0 (–0.44)			
Price of sorghum/kg			1238** (–4.25)			181.4* (–3.11)			
Price of maize/kg					1698* (–3.12)		118.1 (–0.35)	400.3 (–1.28)	
Price of groundnut/kg							–296.9 (–0.83)		–115.8 (–1.24)
Price of millet/kg								45.9 (–0.12)	720.5* (–3.14)
Winter temp.	–887.5* (–2.76)		281.8 (–1.53)	353.8 (–1.08)	423.3 (–1.32)	641.1 (–1.98)	–681.5* (–2.31)	67.7 (–0.3)	–24.2 (–0.04)
Winter temp. sq.	17.8* (–2.55)		–5.2 (–0.91)	–7.6 (–0.80)	–12.1 (–1.2)	–14.5* (–2.28)	25.1* (–3.04)	1.1 (–0.20)	1.9 (–0.10)
Spring temp.	829 (–1.88)		–374 (–1.76)	–338 (–0.88)	–572 (–1.37)	–204 (–0.56)	715 (–1.6)	–50 (–0.16)	–275 (–0.25)
Spring temp. sq.	–15.04 (–1.96)		5.29 (–0.98)	2.9 (–0.29)	12.01 (–1.06)	3.96 (–0.64)	–21.28 (–2.13)	–1.3 (–0.21)	4.16 (–0.22)
Summer temp.	–522 (–1.53)		214 (–1.12)	1207* (–2.24)	458 (–0.93)	–115 (–0.36)	182 (–0.36)	127 (–0.46)	466 (–0.71)

Summer temp. sq.	11.41	−5.07	−23.11*	−2.23	3.45	−0.72	−1.78	−7.36
	(1.95)	(1.41)	(2.09)	(0.19)	(0.7)	(0.07)	(0.36)	(0.68)
Autumn temp.	662*	−320	−1235	728	−2.0	1237	−97	−1.0
	(2.0)	(1.47)	(1.84)	(0.85)	(0.01)	(1.32)	(0.35)	(0.00)
Autumn temp. sq.	−14.86*	8.91	29.94	−26.39	−1.1	−33.99	1.46	−0.67
	(2.47)	(1.98)	(1.99)	(1.20)	(0.20)	(1.54)	(0.26)	(0.06)
Winter prec.	−20.2	4.15	−16.63	27.53	−24.4	−3.47	−0.29	43.99
	(1.98)	(0.53)	(1.29)	(1.80)	(1.96)	(0.28)	(0.02)	(1.52)
Winter prec. sq.	−0.02	−0.05	0.11	−0.07	0.12*	0.05	−0.02	−0.22
	(0.55)	(1.16)	(1.82)	(1.00)	(2.59)	(0.62)	(0.21)	(1.26)
Spring prec.	36.39**	2.92	14.32	−12.17	26.64*	−8.49	−1.66	−7.28
	(3.67)	(0.41)	(1.27)	(1.06)	(2.18)	(0.79)	(0.14)	(0.37)
Spring prec. sq.	−0.07	−0.02	−0.08	0.02	−0.12*	0.02	0.02	−0.12
	(1.15)	(0.57)	(1.93)	(0.56)	(2.02)	(0.49)	(0.45)	(0.37)
Summer prec.	21.0	4.71	4.31	3.65	4.12	−7.64	−3.3	−5.45
	(2.17)	(0.96)	(0.46)	(0.42)	(0.38)	(1.02)	(0.62)	(0.80)
Summer prec. sq.	−0.11*	−0.01	−0.02	0.00	0.02	0.01	0.00	0.03
	(2.55)	(0.64)	(0.66)	(0.12)	(0.47)	(0.35)	(0.10)	(1.11)
Autumn prec.	−36.34**	−15.42*	2.73	5.84	−1.97	8.24	3.4	11.41
	(4.42)	(2.47)	(0.32)	(0.76)	(0.19)	(1.36)	(0.78)	(1.33)
Autumn prec. sq.	0.14**	0.06	0.00	−0.01	−0.04	−0.01	−0.01	−0.06
	(4.36)	(2.62)	(0.01)	(0.33)	(0.86)	(0.57)	(0.75)	(1.55)
Log (area of farmland)	−68.5*	−7.0	−101.4*	−115.8**	−27.6*	−160.0**	−21.9	−72.3*
	(2.86)	(0.29)	(3.08)	(3.48)	(2.23)	(5.68)	(0.71)	(3.27)

Table 10.8 (continued)

Independent variable	Cowpea	Maize	Sorghum	Fruits–vegetables	Maize–beans	Cowpea–sorghum	Maize–groundnut	Maize–millet	Millet–groundnut
	Coef.	Coef.	Coef.	Coef.	Coef.	Coef.	Coef.	Coef.	Coef.
Log (household size)	−14.18		−61.11	33.06	0.38	25.76	178.75*	−37.88	−22.38
	−(0.31)		−(1.36)	−(0.32)	(0.00)	−(0.84)	−(2.36)	−(0.77)	−(0.69)
Dummy household with electricity	−154.6*		170.9	516.1**	280.3*	−15.7	152	112.0	32.6
	−(2.27)		−(1.65)	−(3.81)	−(2.52)	−(0.35)	−(1.34)	−(1.24)	−(0.63)
Soil type 1	188.6		157.6	378.4	−967.6	239.2	1044.4	−247.7	−75.1
	−(0.91)		−(1.31)	−(0.88)	−(1.71)	−(1.35)	−(1.35)	−(0.93)	−(0.35)
Soil type 2	−34.5		232.8*	53.2	146.2	−77.0	107.4	12.7	26.9
	−(0.41)		−(2.83)	−(0.19)	−(0.61)	−(1.86)	−(0.89)	−(0.13)	−(0.23)
Soil type 3	202.9		367.0*	−358.9	−141.4	−193.1	−1583*	119.5	181.7
	−(1.11)		−(2.56)	−(1.45)	−(0.52)	−(0.99)	−(2.04)	−(0.43)	−(0.84)
Soil type 5	117.7		−87.2	−169.7	315.9	332.4	−154.3	412.7	386.0
	−(0.72)		−(0.50)	−(0.48)	−(0.80)	−(1.34)	−(1.10)	−(1.97)	−(0.93)
Selection cowpea			370.8	−385.6	−736.6	111.6	−724.2	313.6	−323.9
			−(1.73)	−(0.60)	−(1.28)	−(1.07)	−(1.48)	−(1.20)	−(1.39)
Selection maize	−160.27		183.4	881.0*	700.8	423.3	323.8	737.5*	753.1
	−(0.93)		−(1.16)	−(2.02)	−(1.34)	−(1.87)	−(0.81)	−(2.73)	−(1.39)
Selection sorghum	509.9			171.3	−247.4	466.5	529.8	−502.0	−476.0
	−(1.37)			−(0.34)	−(0.38)	−(1.12)	−(1.47)	−(1.40)	−(1.02)
Selection fruits–vegetables	372.3		−243.09	0.0	1692.6*	269.1	994.9	−63.7	−713.6
	−(0.91)		−(0.9)	(0.00)	−(2.36)	−(0.85)	−(1.75)	−(0.13)	−(0.64)
Selection maize–beans	−91.4		1338.8*	−285.5		−295.5	−2017.2**	19.9	1424.0
	−(0.20)		−(3.31)	−(0.40)		−(0.63)	−(4.11)	−(0.05)	−(1.00)

190

	(1)	(2)	(3)	(4)	(5)	(6)	(7)	(8)
Selection cowpea–sorghum	−243.1 (−1.44)	−902.4** (−4.57)	309.7 (−0.35)	625.7 (−0.81)		415.1 (−0.89)	−56.6 (−0.27)	61.3 (−0.23)
Selection maize–groundnut	−591.9 (−1.77)	−409.5* (−2.09)	−611.5 (−1.29)	2111.0* (−3.01)	418.7 (−1.14)		−691.0* (−2.57)	−887.7* (−2.30)
Selection maize–millet	996.2* (−2.42)	−128.8 (−0.53)	−1233.0* (−2.07)	272.5 (−0.61)	−572.1 (−1.94)	−682.0 (−1.55)		139.0 (−0.38)
Selection millet–groundnut	−732.6* (−2.05)	−216.0 (−0.94)	1145.9 (−1.5)	−4273.0* (−2.13)	−10.9 (−0.06)	1062.7 (−1.74)	211.2 (−1.06)	
_cons	−1253.4 (−0.74)	3216.6** (−3.51)	−456.5 (−0.13)	−8745.0 (−1.49)	−3274.3 (−1.70)	−15325.0* (−2.47)	−141.8 (−0.08)	−3239.0 (−0.40)
Ancillary								
Sigma^2	822574	1106193	2148307	6178184	318609	3714124	793803	423232
Rho cowpea	−0.23	0.45	−0.34	−0.38	0.25	−0.48	0.45	−0.64
Rho maize	0.72	0.22	0.77	0.36	0.96	0.22	1.06	1.48
Rho sorghum	0.53		0.15	−0.13	1.06	0.35	−0.72	−0.94
Rho fruits–vegetables	−0.13	−0.3		0.87	0.61	0.66	−0.09	−1.41
Rho maize–beans	−0.34	1.63	−0.25		−0.67	−1.34	0.03	2.81
Rho cowpea–sorghum	−0.84	−1.1	0.27	0.32		0.28	−0.08	0.12
Rho maize–groundnut	1.41	−0.5	−0.54	1.09	−0.95		−0.99	−1.75
Rho maize–millet	−1.04	−0.16	−1.08	0.14	−1.3	−0.45		0.27
Rho millet–groundnut		−0.26	1	−2.2	−0.02	0.71	0.3	
Number of obs	367	540	503	381	630	780	313	554
F-test	9.63 (32 334)	29.06 (32 507)	7.8 (31 471)	5.11 (32 348)	7.07 (33 596)	10.49 (33 746)	1.64 (33 279)	2.11 (33 520)

Table 10.8 (continued)

Independent variable	Cowpea	Maize	Sorghum	Fruits–vegetables	Maize–beans	Cowpea–sorghum	Maize–groundnut	Maize–millet	Millet–groundnut
	Coef.	Coef.	Coef.	Coef.	Coef.	Coef.	Coef.	Coef.	Coef.
Prob > F	0.0		0.0	0.0	0.0	0.0	0.0	0.02	0.0
R^2	0.48		0.65	0.34	0.32	0.28	0.32	0.16	0.12
Adjusted R^2	0.43		0.62	0.3	0.26	0.24	0.29	0.06	0.06
Root MSE	268.05		372.05	928.15	582.55	200.22	639.86	349.86	272.07

Notes:
Dependent variable is net revenue. Selection terms are consistent estimators of conditional expected values of the residuals derived from the MNL model in Table 10.2.
* indicates significance at 5% level; ** indicates significance at 1% level.

Source: Kurukulasuriya and Mendelsohn (2008b).

Table 10.9   Marginal effects of climate on conditional net revenue in Africa

Crop	Temperature (°C)	Precipitation (mm/mo)
Cowpea	−5.29*	0.06
Maize		
Sorghum	−19.43*	3.55*
Fruits–vegetables	86.42*	0.03
Maize–beans	−10.40*	14.04*
Cowpea–sorghum	−14.12*	3.85*
Maize–groundnut	31.86*	−5.61*
Maize–millet	3.03*	−3.43*
Millet–groundnut	6.57*	38.19*

Notes:
Marginal effects on the log odds estimated from coefficients in Table 10.8.
* indicates significance at 5% level.

Source: Kurukulasuriya and Mendelsohn (2008b).

a change in temperature and precipitation. Warming will reduce net revenues per hectare for maize–beans, cowpea–sorghum and sorghum but increase net revenues per hectare for maize–groundnut and fruit–vegetables. These results are consistent with the choices made by farmers (warming will increase the probability they will select maize–groundnut and fruit–vegetables).

Table 10.10 presents the change in probability of each crop for the PCM and CCC climate change scenarios. The probability of selecting cowpea decreases under the PCM scenario but it increases under the CCC scenario. Cowpea is a dry weather crop so it is understandable why it increases with the relatively dry CCC scenario and decreases with the relatively wet PCM scenario. The crop combinations maize–millet and millet–groundnut are likely to increase with the PCM scenario but especially with the CCC scenario. Sorghum, maize–beans, maize–groundnut and especially maize are likely to decrease with both scenarios. Fruit–vegetables are likely to increase in the CCC scenario but especially in the PCM scenario. Cowpea is unaffected by the PCM scenario but it is likely to increase with the CCC scenario.

Table 10.11 contains the results of the welfare analysis for the PCM and CCC scenarios. The first row presents the welfare effect of climate change if farmers do not change their crops. With current climate, farmers earn USD157/ha. Without adaptation (growing the exact same crops) and with the PCM climate scenario for 2100, the model predicts revenues to increase by 82 percent. With the CCC scenario, predicted net revenues are

Table 10.10 *Change in African probability of crops under alternative climate scenarios*

	Current climate	PCM	CCC
Cowpea	7.3%	−0.8%	+8.0%
Maize	19.8%	−14.6%*	−15.8%*
Sorghum	10.6%	−6.7%*	−6.6%*
Fruits–vegetables	10.2%	+22.1%*	+8.0%
Maize–beans	7.5%	−6.3%*	−4.1%
Cowpea–sorghum	12.3%	+7.1%	−3.5%
Maize–groundnut	15.2%	−14.3%*	−14.8%*
Maize–millet	6.2%	+7.4%	+5.8%
Millet–groundnut	10.8%	+6.1%	+23.0%*

Notes:
* Indicates significant differences at 5% level.
PCM: Parallel Climate Model; CCC: Canadian Climate Centre model.

Source: Kurukulasuriya and Mendelsohn (2008b).

Table 10.11 *Change in expected welfare in Africa from alternative climate scenarios*

	PCM welfare US$/ hectare	% Change in welfare with PCM	CCC welfare US$/hectare	% Change in welfare with CCC
Crops unchanged	+129	+82%	−106	−68%
Crop switching	+77	+49%	−8	−5%

Notes:
The expected income with current climate is $157/ha.
'Crops unchanged' assumes there is no change in cropping patterns. 'Crop switching' assumes farmers change crops to match the future climate.
PCM: Parallel Climate Model; CCC: Canadian Climate Centre model.

Source: Kurukulasuriya and Mendelsohn (2008b).

likely to fall by 68 percent. The second row calculates the expected welfare effect if farmers are allowed to adjust crop mixes. Welfare increases by 49 percent under the PCM scenario and falls by only 5 percent in the CCC scenario. These results reveal that crop switching is an important adaptation by farmers. Crop switching substantially reduces the estimated impact of climate change.

LIVESTOCK CHOICE

The Structural Ricardian model for livestock in Africa is slightly different from the crop and irrigation models (Seo and Mendelsohn, 2008). The amount of land that farmers use for livestock is not known because African farmers generally rely on common land to graze their animals. Instead of estimating a two-stage model with conditional income per hectare in the second stage, we estimate a three-equation model. The first equation is a selection equation for one of five primary animals: beef cattle, dairy cattle, sheep, goats and chickens. Choice is identified by cross prices. The second equation estimates the size of the herd. This equation is identified by the percentage of grassland in the district. The third equation of the model is a regression of net revenue per animal owned on climate and other variables.

Table 10.12 shows the results of the multinomial logit regression of the probability of choosing each of the five species. The base case is the choice of chickens. Most of the coefficient estimates are very significant. We also report the odds ratio. For example, the odds ratio of electricity for beef cattle is 2.6, which implies that farms with electricity are 2.6 times (odds ratio) more likely to own beef cattle than farms without electricity. Farms with electricity are also more likely to choose sheep, but less likely to choose dairy cattle and goats. Soil variables are weakly significant and their coefficients vary across livestock species. For example, dairy cattle are more likely to be chosen under Cambisol soils, but not under Gleysol soils. With Gleysol soils, beef cattle are more likely to be selected, but not the other species. Livestock price terms are highly significant. Higher milk prices increase the chance that farmers choose dairy cattle and goats. Own animal prices are more difficult to interpret because the price reflects the cost of buying and selling an animal. Higher own prices led farmers away from choosing a species suggesting the prices are acting as input prices. For example, higher beef cattle prices discouraged farmers from selecting beef cattle. Cross-price effects are also evident in Table 10.1. Negative cross-price terms implied that two animals were incompatible, whereas positive cross-price terms imply that they can be raised jointly. For example, the coefficient on the price of goats (chickens) is positive (negative) for beef and dairy cattle, but negative (positive) for sheep, implying that goats (chickens) can be raised with cattle (sheep) but not sheep (cattle).

The climate variables are mostly significant. The quadratic summer temperature coefficients are positive for goats, sheep and dairy cattle, implying a U-shaped function, but the temperature response function for beef cattle is hill-shaped. The quadratic coefficients of the precipitation variables are generally positive, indicating a U-shape except for beef cattle and dairy

Table 10.12 Multinomial logit livestock species selection model for Africa

Variable	Beef cattle			Dairy cattle			Goats			Sheep		
	Coef.	χ^2	Odds ratio	Coef.	χ^2	Odds ratio	Coef.	χ^2	Odds ratio	Coef.	χ^2	Odds ratio
Intercept	5.94	5.58		10.85	45.7		−0.22	0.01		4.92	6.85	
Summer temp.	0.31	2.82	1.37	−0.942	76.5	0.39	−0.168	1.74	0.85	−0.105	0.68	0.90
Summer temp. sq.	−0.0047	1.62	1.00	0.0157	53.3	1.02	0.0049	3.79	1.01	0.0023	0.83	1.00
Summer prec.	0.0058	0.94	1.01	−0.0201	27.6	0.98	−0.0119	8.54	0.99	−0.0155	12.7	0.99
Summer prec. sq.	0.0000	0.22	1.00	0.0001	12.6	1.00	0.0001	14.4	1.00	0.0000	3.03	1.00
Winter temp.	−1.18	73.7	0.31	0.296	5.41	1.35	0.0755	0.16	1.08	−0.435	8.63	0.65
Winter temp. sq.	0.0284	52.6	1.03	−0.0036	1.17	1.00	0.0003	0.00	1.00	0.0139	12.07	1.01
Winter prec.	0.020	5.53	1.02	0.0065	1.44	1.01	−0.0178	8.98	0.98	−0.0243	14.5	0.98
Winter prec. sq.	−0.0001	1.52	1.00	−0.0001	5.98	1.00	0.0001	5.07	1.00	0.0000	0.22	1.00
Soil Cambisols	1.09	1.30	2.97	1.59	7.04	4.93	0.503	0.79	1.65	0.6404	1.33	1.90
Soil Gleysols	−1.86	1.17	0.16	−7.35	25.1	0.00	−2.63	4.23	0.07	−3.67	5.90	0.03
Electricity	0.96	18.3	2.60	−0.078	0.19	0.93	0.0658	0.16	1.07	0.336	4.23	1.40
Beef cattle price	−0.0027	10.06	1.00	0.0002	0.07	1.00	−0.0016	3.32	1.00	−0.0033	16.1	1.00
Milk price	−0.482	11.25	0.62	0.511	24.4	1.67	0.0489	0.20	1.05	−0.0323	0.09	0.97
Goat price	0.0212	5.10	1.02	0.0136	3.92	1.01	−0.0180	3.90	0.98	−0.0261	8.31	0.97
Sheep price	0.0066	1.85	1.01	0.0034	0.79	1.00	−0.0085	2.57	0.99	−0.0037	0.53	1.00
Chicken price	−0.513	14.4	0.60	−1.074	142	0.34	0.416	18.9	1.52	0.765	66.5	2.15

Notes:
The base case is chickens.
The critical value of the Chi-square statistic for the significance at 5% is 3.7. Likelihood ratio test: $P<0.0001$; Lagrange multiplier test: $P<0.0001$; Wald test: $P<0.0001$.

Source: Seo and Mendelsohn (2008).

cattle responses to winter precipitation. Although not shown in Table 10.12, we also test other important control variables such as water availability, altitude and other social statistics such as religions. These variables were dropped since they were not significant. We also tested a number of variables describing the farmer including gender, age and education, but these too were not significant and so were dropped.

The second set of regressions, shown in Table 10.13, predicts the size of the herd. Districts with more natural grassland can support more animals. As reported in Table 10.13, farms in districts with more grassland choose to own more beef cattle, dairy cattle, goats and sheep per household. Farms with electricity own more animals in general except for sheep. Soils are mostly insignificant except for the positive effects of Gleysol soils on the number of sheep. Some of the selection bias correction terms are also significant. Some of the climate variables explain the size of the herd for each species. Summer temperature is significant for dairy cattle, chickens and sheep while winter temperature is significant for goats and sheep. Summer precipitation is significant for beef cattle, chickens and sheep.

Table 10.14 summarizes the results of the third stage, the conditional net revenue per animal regressions. These regressions confirm that the conditional net incomes from the five livestock are sensitive to climate. Seasonal temperature and precipitation explain net revenues. Some soil variables are significant. Gleysol soils are especially harmful to dairy cattle, but beneficial to sheep. The own price of dairy cattle, goats, chickens and sheep has a positive and significant effect while the own price coefficient of beef cattle is insignificant. The selection bias coefficients reveal interactions among the species; for example, the selection term coefficient for sheep is positive in the beef cattle income regression. Farmers who the selection model predicted would choose sheep, but who actually choose beef cattle, have higher than expected beef incomes. Whereas if the selection model predicts the farmer would have chosen chickens but the farmer actually chose beef cattle, that farmer would have lower beef cattle income.

Table 10.15 shows how all three decisions change with a marginal change in climate. As temperature rises, farmers switch away from choosing beef cattle, dairy cattle and chickens and towards goats and sheep. Farmers reduce the size of their beef cattle herd but increase the number of their goats and sheep. These changes are supported by the marginal changes in net revenue per animal. Warmer temperatures decrease beef cattle income but increase sheep income. As rainfall increases, farmers move away from beef cattle, dairy cattle and sheep towards goats and chickens. With the increase in rainfall, the net income of beef cattle, dairy cattle and sheep declines, but the net income of goats and chickens increases. Although not

Table 10.13   Conditional herd size by species regressions in Africa

Variable	Beef cattle		Dairy cattle		Chickens		Goats		Sheep	
	Coef.	t-stat.	Coef.	t-stat.	Coef.	t-stat.	Coef.	t-stat.	Coef.	t-stat.
Intercept	223	0.82	-58.05	-3.20	1981	1.80	20.8	0.65	14.5	0.70
Summer temp.	-12.6	-0.52	3.78	2.90	-160	-2.08	2.35	1.28	7.59	4.31
Summer temp. sq.	0.229	0.51	-0.063	-2.77	3.66	2.42	-0.037	-1.07	-0.123	-3.81
Winter temp.	-25.5	-1.09	1.21	0.83	0.046	0.00	-5.55	-1.96	-10.6	-4.34
Winter temp. sq.	0.488	0.72	-0.055	-1.43	0.143	0.06	0.142	2.12	0.224	3.65
Summer prec.	2.13	3.14	0.089	1.76	-2.42	-1.20	-0.042	-0.88	0.212	4.36
Summer prec. sq.	-0.01	-3.36	0.000	-1.89	0.022	2.85	0.000	0.71	0.000	-0.66
Winter prec.	-0.378	-0.39	0.204	3.10	1.96	0.80	0.010	0.13	0.148	1.69
Winter prec. sq.	-0.002	-0.29	-0.001	-2.45	-0.008	-0.49	0.000	-0.03	0.002	2.43
Soil Cambisols	8.87	0.09	-4.31	-0.77	251	1.05	0.806	0.21	-2.33	-0.73
Soil Gleysols	4.32	0.02	12.6	0.73	-436	-0.89	-26.8	-1.47	62.4	3.17
Electricity dummy	145	5.35	0.665	0.36	298	3.59	6.04	3.51	-1.33	-0.71
% grasslands	261	3.41	14.1	1.72	-208	-0.92	3.91	0.70	41.5	6.37
Cattle beef – selection			-4.73	-0.75	219	0.63	29.7	3.21	12.5	1.19
Cattle dairy – selection	3.71	0.08			-132	-0.70	-5.34	-0.94	-35.2	-4.98
Goats – selection	-546	-3.14	-5.45	-0.48	497	0.89			17.9	1.59
Sheep – selection	426	3.74	-16.5	-1.58	-325	-0.79	1.51	0.24		
Chickens – selection	86.1	0.72	23.3	3.44			-25.9	-2.53	16.6	1.67
Adjusted R²	0.33		0.12		0.12		0.03		0.09	
N	381		1036		876		774		810	

Source: Seo and Mendelsohn (2008).

Table 10.14   Conditional net revenue per animal regressions in Africa

Variable	Beef cattle		Dairy cattle		Chickens		Goats		Sheep	
	Estimate	t-statistic	Estimate	t-statistic	Estimate	t-statistic	Estimate	t-statistic	Estimate	t-statistic
Intercept	280	0.90	6.83	0.04	3.92	1.19	-17.1	-0.45	69.3	2.82
Summer temp.	63.3	2.35	-7.98	-0.53	0.621	2.66	-5.35	-2.04	3.47	1.84
Summer temp. sq.	-1.15	-2.37	0.300	1.11	-0.010	-2.16	0.119	2.54	-0.041	-1.16
Winter temp.	-143	-6.22	0.639	0.04	-1.29	-4.81	7.13	1.88	-10.04	-3.72
Winter temp. sq.	3.94	6.26	0.129	0.32	0.031	4.33	-0.196	-2.27	0.21	2.93
Summer prec.	-2.76	-3.71	-0.390	-0.78	0.007	1.08	0.033	0.56	-0.068	-1.30
Summer prec. sq.	0.009	2.74	0.002	0.84	0.000	2.32	0.000	0.77	0.000	0.92
Winter prec.	-0.882	-0.76	-0.409	-0.66	0.006	0.78	0.028	0.26	0.004	0.04
Winter prec. sq.	-0.003	-0.46	0.002	0.64	0.000	3.36	0.000	0.73	0.000	0.03
Soil Cambisols	-78.4	-0.70	92.9	1.80	-0.789	-1.10	-1.24	-0.27	-3.33	-0.90
Soil Gleysols	-428	-1.90	-516	-3.08	0.680	0.46	-18.07	-0.82	58.1	2.67
Electricity	147	4.96	-24.08	-1.35	0.376	1.87	2.13	1.01	8.18	3.94
Own price	-44.1	-1.67	27.2	2.39	0.419	4.00	4.82	3.34	37.7	4.76
Cattle beef – selection	-61.2	-1.22	-139	-2.01	3.25	3.60	4.82	3.34	37.7	4.76
Cattle dairy – selection					-1.83	-3.49	-8.04	-0.54	51.2	4.75
Goats – selection	205	0.99	380	3.39	5.65	3.55			-6.09	-0.48
Sheep – selection	313	2.47	232	2.29	-6.33	-5.14	-16.7	-1.80		
Chickens – selection	-562	-4.13	-489	-7.07			5.38	0.44	-19.2	-1.76
Adjusted R²	0.75		0.27		0.14		0.17		0.20	
N	333		1043		888		775		842	

Source: Seo and Mendelsohn (2008).

Table 10.15   Marginal climate effects on probability of selection amongst African livestock

	Beef cattle	Dairy cattle	Goats	Sheep	Chickens
Probability (%)					
Baseline	9.40	27.3	20.8	21.8	22.9
Temperature	−1.29	−1.34	+1.09	+1.68	−0.84
Precipitation	+0.2	−0.02	−0.01	−0.36	+0.22
Herd size (head/farm)					
Baseline	57.4	57.4	57.4	57.4	1.55
Temperature	−7.99	−7.99	−7.99	−7.99	+0.07
Precipitation	−0.32	−0.32	−0.32	−0.32	+0.03
Net Revenue ($/head)					
Baseline	221	145	11.5	17.8	1.55
Temperature	−5.57	+11.07	−0.42	+0.13	+0.07
Precipitation	−0.72	−0.01	+0.08	−0.02	+0.03

Note: Temperature is measured in °C and precipitation in mm/month.

Source: Seo and Mendelsohn (2008).

completely consistent across Tables 10.12, 10.13 and 10.14, the three analyses combined in Table 10.15 suggest that farmers would substitute beef cattle for goats and sheep as temperature rises and they would substitute beef cattle and sheep for goats and chickens as rainfall increases.

Table 10.16 shows how the PCM and CCC scenarios will be likely to affect livestock in Africa. We rely on the parameters estimated in Table 10.12 to predict how the species selection may change. Under the wet PCM scenario, beef cattle, dairy cattle, sheep and chickens all decline in favor of goats. Under the CCC scenario, farmers will shift away from dairy cattle, goats and chickens to sheep. The coefficients in Table 10.13 are used to predict how herd size will change. With the PCM scenario, beef cattle will fall precipitously and goats, sheep, and chicken will all increase. With the CCC scenario, sheep will not increase as much as in the PCM scenario but there will be an even larger increase in goats. Finally, the coefficients in Table 10.14 are used to predict how net revenue will change. With the PCM scenario, beef cattle and sheep income per head fall sharply, but dairy cattle and especially chicken incomes per head increase. With the CCC scenario, beef cattle, dairy cattle and chicken incomes increase, but goat incomes fall sharply.

The change in expected net income can be calculated by combining the results from Table 10.16 across species. Multiplying the probability of

Table 10.16   Predicted African livestock changes from climate scenarios

	Beef cattle	Dairy cattle	Goats	Sheep	Chickens
Species probability (%)					
Baseline	9.40	27.3	20.8	21.8	22.9
PCM	−1.83	−7.35	+10.7	−2.19	−4.55
CCC	−0.83	−6.53	−3.34	+15.9	−9.94
Herd size (head/farm)					
Baseline	57.4	5.65	11.3	13.07	137
PCM	−30.4	−0.59	+8.95	+34.1	+178
CCC	−27.6	−1.39	+24.8	+9.97	+204
Net revenue ($/head)					
Baseline	224	150.1	11.1	16.7	1.60
PCM	−66.8	+44.8	+1.90	−4.12	+3.46
CCC	+100.6	+68.1	−7.84	+1.69	+1.82

Source: Seo and Mendelsohn (2008).

*Table 10.17   Predicted impacts of climate change in 2100 on African
livestock expected income*

Expected income	(US$/farm)	% change	Total (billions US$)
Current	882		60.0
PCM	+55.2	+6%	+63.6
	(−24.8, +135)	(−3%, +15%)	(58.2, 69.0)
CCC	+1488	+169%	+161.4
	(+1343, +1633)	(+152%, +186%)	(151.2, 171.6)

Note: 95% confidence intervals in parenthesis generated with bootstrap method.

Source: Seo and Mendelsohn (2008).

each species by the size of the herd multiplied by the net income per animal
gives the expected income of that animal. Summing these values across
animals gives the total expected income from livestock seen in Table 10.17.
The 95 percent confidence interval was calculated using 200 bootstrap
runs. Expected income is different from actual income even given current
climate. With expected income, each farmer is assigned a probability of
owning an animal given what the researcher knows about each farmer.
With actual income, each farmer has selected an animal given what the

farmer knows. The current expected income from livestock management is almost USD900 per farm. With the PCM scenario, this increases
about USD50 (6 percent) per farm but the increase is not significant.
With the CCC scenario, expected income increases a dramatic USD1500
(169 percent). With the CCC scenario, livestock income in Africa could
increase by a USD100 billion.

FARM TYPE CHOICE

The farm type study in South America identifies five types of farms: crop-
only rainfed, crop-only irrigated, mixed (crops and livestock) rainfed,
mixed irrigated, and livestock-only farms. The analysis measures the
choice of each farm type and then the conditional land value for each type
of farm.

The choice of farm type is estimated using a multinomial logit with
livestock-only farms as the base case. The four regressions are presented
in Table 10.18. The climate coefficients are mostly very significant. For
the four regressions, the response to higher temperatures is concave
whereas the response to higher precipitations is convex. Acrisol, Cambisol
and Gleysol soils decrease the probability of each farm type relative to
livestock-only farms. Planosol soils, however, increase the probability of
crop-only rainfed and mixed rainfed farming being chosen. Commercial
farms do not show substantially different responses. Crop-only irrigated
farms are more often chosen when tomato prices are high. When maize
prices are high, each of the four farm types are chosen more often compared to livestock-only. However, when potato prices are high, livestock-
only is chosen more often. Table 10.19 describes the marginal effects of
temperature and precipitation on farm type choice. Higher temperatures
encourage farmers to move away from crop-only farms to mixed farms
and livestock farms. Higher precipitation pushes farmers to adopt rainfed
farming and avoid expensive irrigation investments.

The second stage estimates of conditional land values for each farm
type are displayed in Table 10.20. The climate coefficients are significant
and differ in magnitude and sign by farm type. For example, summer
temperature is significant for rainfed crop-only farms and livestock-only
farms but crop-only farms have a concave response to summer temperature, whereas livestock farms have a convex response. Summer precipitation is highly significant for all types of farms. Farmland values have a
concave response to summer precipitation for all farm types except mixed
irrigated farms. Soils play an important role in determining land value.
Gleysol soils reduce farmland values, and Luvisol soils increase farm land

Table 10.18   Multinomial logit model of farm type selection in Latin American Structural Ricardian model

Variables	Crop-only rainfed		Crop-only irrigated		Mixed rainfed		Mixed irrigated	
	Coef.	χ^2	Coef.	χ^2	Coef.	χ^2	Coef.	χ^2
Intercept	-5.9049	5.86	-1.0095	0.18	-5.7757	6.28	-2.269	0.85
Summer temp.	0.7091	8.44	0.5879	5.7	0.6647	8.63	0.2141	0.76
Summer temp. sq.	-0.0312	23.3	-0.0264	15.99	-0.0259	19.48	-0.0148	5.14
Summer prec.	-0.0235	15.67	-0.0375	38.55	-0.0215	14.47	-0.0358	30.88
Summer prec. sq.	0.000051	11.19	0.000071	19.62	0.000044	9.03	0.000075	20.15
Winter temp.	1.1373	88.69	0.5756	21.79	0.9017	70.53	1.1156	64.44
Winter temp. sq.	-0.0277	55.65	-0.0115	8.58	-0.0196	36.49	-0.0273	40.08
Winter prec.	-0.0236	16.88	-0.0282	20.66	-0.0164	8.9	-0.024	13.93
Winter prec. sq.	0.000099	12.04	0.000083	7.12	0.000079	8.13	0.000069	4.62
Soil Acrisols	-0.0206	11.59	-0.0363	12.63	-0.0145	8.75	-0.0383	9.47
Soil Cambisols	-0.0161	1.68	-0.0454	8.89	-0.00015	0	-0.0133	0.95
Soil Gleysols	-0.0127	1.65	-0.0629	8.25	-0.0222	5.52	-0.0335	4.83
Soil Andosols	-0.00088	0.01	-0.0597	9.72	0.0106	0.86	-0.026	2.7
Soil Planosols	0.0061	2.67	-0.0125	6.26	0.00706	4.42	0.00463	1.1
Soil Xerosols	-0.0073	0.42	-0.0246	4.3	0.00145	0.02	-0.0113	0.87
Maize price	1.1629	24.5	1.079	20.84	1.3226	32.58	1.4419	35.08
Tomato price	-6.0386	29.03	-0.2632	0.17	-5.6558	34.21	-6.5195	16.63
Altitude	-0.00008	0.06	0.00056	2.89	0.000592	3.72	0.000309	0.74
Sand and clay	-0.2096	1.89	-0.4702	6.63	-0.1984	2.14	0.0637	0.09
Clay	-0.0876	0.35	0.0238	0.02	-0.0263	0.04	0.1924	1.2
Steep	-0.16	0.73	-0.294	2.88	-0.2057	1.48	-0.6127	10.8

Notes: Omitted choice is livestock only.
Three test statistics and P-values for the model significance are Likelihood Ratio test: P<0.0001; Lagrange Multiplier test: P<0.0001; Wald test: P<0.0001.

Source: Mendelsohn and Seo (2007).

Table 10.19   *Marginal climate effects on the probability of farm types in Latin America*

	Crop-only rainfed	Crop-only irrigated	Mixed rainfed	Mixed irrigated	Livestock-only
Baseline	17.99%	15.99%	47.41%	8.99%	9.63%
Temperature (°C)	−1.75%	−1.83%	+1.93%	+0.39%	+1.26%
Precipitation (mm/mo)	+0.03%	−0.14%	+0.09%	−0.06%	+0.09%

Source: Mendelsohn and Seo (2007).

values. Planosol soils increase the land value of crop-only rainfed farms, but decrease the value of mixed rainfed farms. Clay soils generally reduce agricultural land values with the exception of crop-only irrigated farms. Table 10.20 also displays the selection correction terms for each farm type. Many of the selection terms are significant, which suggests that selection bias is important in this data set. For example, when the selection model suggests a farm should be mixed irrigated, that farm has a lower value if it is actually crop-only rainfed or livestock-only. When the selection model predicts a farm is livestock-only, that farm has a lower value if it is actually crop-only irrigated or mixed rainfed. On the other hand, if the selection model predicts a farm will be crop-only, it has a higher value if it is actually a mixed farm.

Table 10.21 presents the marginal effects and elasticities of annual temperature and precipitation evaluated at the mean of the sample with respect to conditional land value. Warming decreases the value of crop-only farms and especially livestock-only farms but it increases the value of mixed farms. Increased precipitation raises the value of all farms but especially livestock-only and crop-only farms.

Table 10.22 examines how the PCM and CCC climate scenarios change the probabilities and conditional land values of South American farms. With the mild and wet PCM scenario, there will be more mixed farms and slightly more crop-only rainfed farms but a drop in livestock-only farms. Growing crops in this scenario is slightly more attractive. Further, irrigation is less important with the increase in precipitation. In the CCC scenario, there will be fewer crop-only farms and fewer mixed farms. In contrast, there will be a dramatic increase in livestock-only farms. Table 10.22 also presents the changes in conditional land value for each climate scenario. The results are consistent with the choice decisions. With the PCM scenario, rainfed crops earn substantially more, but the income

Table 10.20   Conditional income model in Latin American Structural Ricardian model

Variable	Crop-only rainfed		Crop-only irrigated		Livestock-only		Mixed rainfed		Mixed irrigated	
	Coef.	t-stat.	Coef.	t-stat.	Coef.	t-stat.	Coef.	t-stat.	Coef.	t-stat.
Intercept	-5384.7	-1.26	2127.2	0.54	25166.0	3.19	-3358.0	-1.81	-19947.0	-3.84
Summer temp.	866.9	2.53	491.6	1.21	-2078.5	-2.73	207.8	1.33	882.3	2.05
Summer temp. sq.	-30.40	-2.88	-6.20	-0.5	47.16	2.18	-8.19	-1.77	-31.09	-2.36
Winter temp.	167.0	0.57	-1161.7	-3.96	184.3	0.37	502.1	5.54	2201.6	4.86
Winter temp. sq.	-5.91	-0.7	29.84	3.26	4.93	0.36	-19.24	-7.07	-61.96	-5.04
Summer prec.	6.69	0.86	60.71	3.16	-14.57	-0.66	29.32	6.66	-17.67	-1.03
Summer prec. sq.	-0.02	-1.28	-0.14	-2.88	0.02	0.33	-0.06	-6.03	0.06	1.61
Winter prec.	-20.87	-3.38	53.63	3.48	0.17	0.01	22.56	4.24	7.13	0.48
Winter prec. sq.	0.07	3.76	-0.20	-2.88	-0.09	-1.48	-0.07	-2.59	-0.04	-0.74
Soil Acrisols	16.79	1.56	52.32	1.05	0.44	0.04	4.76	0.92	-36.49	-1.22
Soil Cambisols	-8.11	-0.44	102.69	2.41	-0.04	0	4.93	0.69	68.62	2
Soil Gleysols	0.22	0.01	-23.87	-0.29	-130.41	-4.23	30.20	4.06	38.56	1.08
Soil Andosols	7.63	0.57	75.67	1.15	-13.14	-0.58	26.75	3.86	-12.37	-0.27
Soil Planosols	9.85	1.37	-6.73	-0.47	-16.14	-2.3	9.71	3.06	21.37	1.7
Soil Xerosols	-1.87	-0.1	-2.51	-0.09	-22.96	-1.2	7.65	0.94	34.68	1.43
Maize price	-202.5	-0.81	-973.2	-4.77	3671.2	3.4	-464.0	-5.24	881.5	2.75
Tomato price	-3205.4	-0.9	-2715.3	-2.45	-2105.5	-1.73	-1304.1	-1.2	-22292.0	-4.71
Electricity dummy	70.5	0.17	-176.4	-0.21	-31.1	-0.12	307.8	1.55	596.2	0.95
Log household size	221.2	0.84	338.5	0.94	211.0	0.9	-47.0	-0.36	-46.2	-0.16
Log education	105.9	0.53	-553.2	-1.78	-532.3	-2.33	-135.7	-1.52	220.8	0.97
Female dummy	-546.5	-1.14	-276.5	-0.44	-83.7	-0.19	89.6	0.39	-900.1	-1.56

Table 10.20 (continued)

Variable	Crop-only rainfed		Crop-only irrigated		Livestock-only		Mixed rainfed		Mixed irrigated	
	Coef.	t-stat.	Coef.	t-stat.	Coef.	t-stat.	Coef.	t-stat.	Coef.	t-stat.
Select crop-only rainfed			-3566.3	-1.21	-14536.0	-3.73	5432.5	5.42	7765.9	2.88
Select crop-only irrigated	1575.9	0.83			2312.2	1.13	-4197.2	-5.01	-7205.7	-2.38
Select mixed rainfed	-1476.4	-0.77	15389.0	3.66	5658.9	1.81			1371.1	0.38
Select mixed irrigated	-1123.8	-0.51	-9616.1	-3.42	6891.7	2.04	-3275.9	-2.77		
Select livestock only	170.0	0.12	-2379.0	-3.98			2611.0	3.7	-3711.8	-1.39
N	365		179		182		949		268	
Adjusted R²	0.18		0.17		0.17		0.30		0.29	

Source: Mendelsohn and Seo (2007).

Table 10.21   Marginal climate effects and elasticities on conditional incomes in Latin America

Farm type	Temperature (°C)	Precipitation (mm/mo)	Temperature	Precipitation
	Marginal effects		Elasticities	
Rainfed crop-only	−293.6	−8.15	−2.09	−0.47
Irrigated crop-only	−266.0	66.92	−1.22	1.56
Crop/livestock rainfed	19.6	33.30	0.09	0.78
Crop/livestock irrigated	483.9	−4.95	4.31	−0.24
Livestock-only	216.3	−20.32	3.04	−1.66

Source: Mendelsohn and Seo (2007).

Table 10.22   Impacts on each farm type of climate scenarios in Latin American Structural Ricardian model

	Crop-only rainfed	Crop-only irrigated	Mixed rainfed	Mixed irrigated	Livestock-only
Probabilities					
Baseline	17.99%	15.99%	47.41%	8.99%	9.63%
PCM	+1.82%	−1.14%	+2.35%	+1.32%	−4.35%
CCC	−3.51%	−2.54%	+1.08%	+0.11%	+4.87%
Conditional income					
Baseline	2329	2809	1561	1673	1229
PCM	+515	−1497	−540	+1167	+51
CCC	−2324	−768	−1958	−1098	+3195

Note: Calculated from coefficients in Tables 10.20 and 10.21.

Source: Mendelsohn and Seo (2007).

from irrigated crops falls. The increased precipitation makes installing costly irrigation unattractive. As temperatures increase, income from livestock management also increases, but less than with crops. Farmers consequently shift away from livestock-only farms towards crops-only and mixed farms. With the CCC scenario, income from growing crops alone in rainfed farms falls precipitously, whereas incomes from crop-only irrigated, mixed rainfed, and livestock-only farms are more resilient. The selection model consequently predicts a reduction in crop-only farms. Curiously, the selection model also predicts a large shift from crop-only irrigated farms to livestock-only. This shift is not supported by the change in conditional land values.

Table 10.23   Climate change impacts on expected land value

Baseline	Climate change impacts ($/ha)	% change
PCM	1876	
	−150	−8.0
	(−356, +56)	(−19.0, 3.0)
CCC	+513	+27.3
	(+345, +681)	(+18.4, +36.3)

Note: 95% confidence intervals are in parenthesis.

Source: Mendelsohn and Seo (2007).

Combining the changes in probability and the change in land value, one can estimate the changes in expected land values. Table 10.23 reports the current expected land value and the change under each climate scenario. With the PCM scenario, the structural model predicts that the expected value of farmland will increase in South America by almost 50 percent. Most of these gains come from the huge increase in value of rainfed crop-only farms. Despite the adaptations that farmers make to climate changes under the CCC scenario, expected land values are expected to fall in that scenario by 17 percent. Most of these losses come from damages to crop-only rainfed farms.

REFERENCES

Basist, A., C. Williams, N. Grody, T. Ross, S. Shen, A. Chang, R. Ferraro and M. Menne (2001), 'Using the special sensor microwave imager to monitor surface wetness', *Journal of Hydrometeorology*, **2**, 297–308.

Boer, G., G. Flato and D. Ramsden (2000), 'A transient climate change simulation with greenhouse gas and aerosol forcing: projected climate for the 21st century', *Climate Dynamics*, **16**, 427–50.

Caswell, M. and D. Zilberman (1986), 'The effect of well depth and land quality on the choice of irrigation technology', *American Journal of Agricultural Economics*, **68**, 798–811.

Dinar, A. and D. Yaron (1990), 'Influence of quality and scarcity of inputs on the adoption of modern irrigation technologies', *Western Journal of Agricultural Economics*, **15**(2), 224–33.

Dinar, A. and D. Zilberman (1991), 'The economics of resource-conservation, pollution-reduction technology selection: the case of irrigation water', *Resources and Energy*, **13**, 323–48.

Dinar, A., M.B. Campbell and D. Zilberman (1992), 'Adoption of improved irrigation and drainage reduction technologies under limiting environmental conditions', *Environmental and Resource Economics*, **2**, 373–98.

Dinar, A., R. Hassan, R. Mendelsohn and J. Benhin (2008), *Climate Change and Agriculture in Africa: Impact Assessment and Adaptation Strategies*, London: Earthscan.

Emori, S., T. Nozawa, A. Abe-Ouchi, A. Namaguti and M. Kimoto (1999), 'Coupled ocean-atmospheric model experiments of future climate change with an explicit representation of sulfate aerosol scattering', *Journal of the Meteorological Society of Japan*, **77**, 1299–1307.

FAO (Food and Agriculture Organization) (1997), *Irrigation Potential in Africa: A Basin Approach*, FAO Land and Water Bulletin, 4, FAO Land and Water Development Division, Rome, Italy.

FAO (Food and Agriculture Organization) (2003), *The Digital Soil Map of the World*: Version 3.6 (January), Rome, Italy.

Heckman, J. (1979), 'Sample selection bias as a specification error', *Econometrica*, **47**, 153–61.

IPCC (Intergovernmental Panel on Climate Change) (2007), *State of the Science*, Cambridge: Cambridge University Press.

IWMI (International Water Management Institute) and The University of Colorado (2003), 'Hydroclimatic Data. Climate, water and agriculture: impacts and adaptation of agro-ecological systems in Africa', Discussion Paper 13, CEEPA, University of Pretoria, South Africa.

Kurukulasuriya, P. and M.I. Ajwad (2007), 'Application of the Ricardian technique to estimate the impact of climate change on smallholder farming in Sri Lanka', *Climatic Change*, **81**, 39–59.

Kurukulasuriya, P. and R. Mendelsohn (2008a), 'Modeling endogenous irrigation: the impact of climate change on farmers in Africa', World Bank Policy Research Working Paper 4278, Washington, DC, USA: World Bank.

Kurukulasuriya, P. and R. Mendelsohn (2008b), 'Crop switching as an adaptation strategy to climate change', *African Journal of Agriculture and Resource Economics*, **2**, 105–26.

Kurukulasuriya, P., R. Mendelsohn, R. Hassan, J. Benhin, M. Diop, H.M. Eid, K.Y. Fosu, G. Gbetibouo, S. Jain, A. Mahamadou, S. El-Marsafawy, S. Ouda, M. Ouedraogo, I. Sène, S.N. Seo, D. Maddison and A. Dinar (2006), 'Will African agriculture survive climate change?', *World Bank Economic Review*, **20**, 367–88.

Mendelsohn, R. and A. Dinar (2003), 'Climate, water, and agriculture', *Land Economics*, **79**, 328–41.

Mendelsohn, R. and S.N. Seo (2007), 'An integrated farm model of crops and livestock: modeling Latin American agricultural impacts and adaptation to climate change', World Bank Policy Research Working Paper 4161, Washington, DC, USA: World Bank.

Mendelsohn, R., A.F. Avila and S.N. Seo (2007), Incorporation of climate change into rural development strategies: synthesis of the Latin American results, Montevideo, Uruguay: PROCISUR.

Mendelsohn R., W. Nordhaus and D. Shaw (1994), 'The impact of global warming on agriculture: a Ricardian analysis', *American Economic Review*, **84**, 753–71.

Mendelsohn, R., A. Basist, A. Dinar, F. Kogan, P. Kurukulasuriya and C. Williams (2006), 'Climate analysis with satellites versus weather station data', *Climatic Change*, **81**, 71–83.

Negri, D.H. and D.H. Brooks (1990), 'Determinants of irrigation technology choice', *Western Journal of Agricultural Economics*, **15**, 213–23.

Schlenker, W., M. Hanemann and A. Fischer (2005), 'Will US agriculture really benefit from global warming? Accounting for irrigation in the hedonic approach', *American Economic Review*, **95**(1), 395–406.

Seo, S.N. and R. Mendelsohn (2008), 'Measuring impacts and adaptation to climate change: a structural Ricardian model of African livestock management', *Agricultural Economics*, **38**, 150–65.

USGS (United States Geological Survey) (2004), 'Global 30 Arc Second Elevation Data', data set, USGS National Mapping Division, EROS Data Centre.

Washington, W., J. Weatherly, G. Meehl, A. Semtner, B. Bettge, A. Craig, W. Strand, J. Arblaster, V. Wayland, R. James and Y. Zhang (2000), 'Parallel Climate Model (PCM): control and transient scenarios', *Climate Dynamics*, **16**, 755–74.

Weng, F. and N. Grody (1998), 'Physical retrieval of land surface temperature using the Special Sensor Microwave Imager', *Journal of Geophysical Research*, **103**, 8839–48.

World Bank (2003), *Africa Rainfall and Temperature Evaluation System (ARTES)*, Washington, DC, USA: World Bank.

11. Summary of results

Agronomic research on crop productivity in controlled settings suggests that most individual crops have well defined climate ranges where they perform best (Chapter 2; Reilly et al., 1996; IPCC, 2007b). As one moves out of these ranges, either to cooler or warmer temperatures or to dryer or wetter conditions, crop productivity falls. The ideal climate zone for each crop is quite different. This is why one observes that specific crops tend to be grown in certain places and not others.

Chapter 3 reviews a number of methodologies, used in the literature, to measure climate impacts on agricultural crops. Some of these methods ignore adaptation. They assume, for example, that crops continue to be grown where they are grown today. Even small changes in climate can have deleterious effects on crop productivity in this case. The methods that ignore adaptation tend to have more pessimistic results. For example, the predictions of the agronomic literature tend to be more dire (Rosenzweig and Parry, 1994; Parry et al., 2005). This literature tends to predict that climate change will lead to substantial reductions in yields.

Economic research that takes into account adaptation, however, suggests that crop choice and livestock choice are endogenous. Farmers select which crop to grow given what is most profitable. Economic models of farm behavior challenge the assumption at the heart of the agronomic studies, that crop choice and livestock choice are exogenous and unchanging. Instead, economists find that farmers will switch crops and livestock in direct response to climate change. The outcome of climate change depends on adaptation. Climate research on agricultural impacts must take farm adaptation into account or it will overestimate damages.

Of course, just because farmers can switch crops or livestock does not mean that they will not be affected by climate change. The farmer will earn an initial profit from growing the ideal crop or animal for current climate conditions. The farmer will earn a different profit from growing a new ideal crop or animal in future climate conditions. The magnitude of the welfare change is the change in profits. The change is a loss if the farmer could earn more in the initial climate, and it is a benefit if the farmer can earn more in the new climate.

Chapters 4, 5 and 6 examine some promising tools for using cross-sectional analysis to measure climate impacts and adaptation.

Cross-sectional analysis is an excellent tool for studying comparative static equilibriums. Both livestock and crops can be studied with these techniques. By analyzing the net revenue or land value of a farm in one climate versus another, it is possible to measure long-term climate sensitivity. Because farmers and ecosystems adapt to the climate they are in, the measurements account for adaptation and indirect ecological effects. The traditional Ricardian method is therefore an ideal tool for measuring climate impacts. The cross-sectional analysis of farm choices also reveals how farmers will adapt to climate change in the long run. Finally, the structural Ricardian models combine the insights of the adaptation choice models with the income effects measured by the Ricardian model. These last models provide not only a sense of what farmers may do but also what incomes they will earn. However, the cross-sectional studies do not measure dynamic changes. The studies discussed in Chapters 4, 5 and 6 do not take into account transition costs associated with moving from one steady state to another.

IMPACTS

As discussed in Chapter 7, the traditional Ricardian models have been applied to many countries. There have been empirical studies in the United States, Canada, Germany, England, India, Sri Lanka, Israel, 11 countries in Africa, 7 countries in South America, and China. The results reveal that there are important seasonal effects of climate. The off-season as well as the growing-season appears to be important. Off-season effects can be important because they affect weeds and pests, which are important to farmer net revenues. Seasonal effects are important because plants react to temperature and precipitation differently depending upon the stage of their life cycle. The results often find that marginal temperature is harmful in one season and beneficial in the next. Because of these offsetting seasonal effects, annual effects are often not as dramatic as seasonal effects. Finally, seasonal responses are non-linear, some being concave and some being convex. Although the agronomic literature generally suggests that at least annual temperature effects are convex, it is not clear there is an a priori hypothesis about the shape of seasonal responses.

In order to get a sense of the magnitude of results from different studies, we examine two climate scenarios: a moderate scenario (PCM) and a harsh scenario (CCC or HAD3). A Ricardian model for the United States suggests that the PCM scenario would lead to a small increase in net income that would be slightly beneficial. The more severe CCC scenario would lead to no net change. In both cases, benefits in some regions offset damages in another.

A different study that takes surface water withdrawals into account supports these results (Mendelsohn and Dinar, 2003). However, if one examines only counties that depend on rainfed farming, the results are mildly harmful for the US (Schlenker et al., 2005). This last result suggests that rainfed farms are more climate-sensitive than irrigated farms in the US.

Analyses of India and Brazil suggest that the agricultural sectors in these two countries are more sensitive to climate change than the US (Mendelsohn and Dinar, 1999; Mendelsohn et al., 2001; Sanghi and Mendelsohn, 2008). With a 2°C warming and an 8 percent increase in precipitation, Indian net revenue will fall 12 percent and Brazilian farmland values will fall 20 percent. With a 3.5°C warming and no increase in precipitation, Indian net revenue will fall 21 percent and Brazilian farmland values will fall 29 percent. Brazilian climate sensitivities may be slightly higher than Indian sensitivities because of the greater reliance on irrigation in Indian agriculture. Both Brazil and India are more sensitive than the US, because they are already in hotter climates.

Extending these Ricardian results to Africa reveals that African farmers are also vulnerable to climate change (Kurukulasuriya and Mendelsohn, 2008a). With the PCM climate scenario, the crop net revenue of African farmers would increase 48 percent. But with the 5+°C warming of the CCC scenario, farmers would lose –33 percent of their crop net revenue. Rainfed farmers are even more vulnerable. With the PCM scenario, they might gain 50 percent more crop net revenue, but with the CCC scenario, they would lose –43 percent of their crop net revenue. In contrast, irrigated farmers were much less sensitive. In the PCM scenario, they would gain 30 percent more crop net revenue, and in the CCC scenario they would gain 5 percent more crop net revenue provided they still had water to irrigate. Note that all of these Ricardian results assume that farmers can adapt using current available technology.

Another important result from Africa is that one must also measure climate impacts on livestock (Seo and Mendelsohn, 2008a). Livestock farms were divided into small and large on the basis of the size of their herds. The large farms were found to be climate-sensitive whereas the small farms were robust. With the PCM scenario, small farms were expected to increase livestock net revenue by 54 percent, whereas livestock net revenue in large farms was expected to fall –23 percent. With the CCC scenario, livestock net income was expected to rise 323 percent in small farms but fall by 26 percent in large farms. The reliance on heat intolerant beef cattle in the large livestock farms makes them more vulnerable than the small livestock farms that use a much wider set of animals. At least for small farmers, the increase in livestock income helps compensate for the losses in crop income.

Another set of Ricardian results concerns farmland values in Latin America (Seo and Mendelsohn, 2008b). These results reflect both crop and livestock net effects. The farms were divided into small (household) and large (commercial) farms. With the mild PCM scenario, small farms suffer an 11 percent loss in farmland value and large farms a 20 percent loss. With the CCC scenario, small farms suffer a 49 percent loss of farm-land value and large farms suffer a 69 percent loss. Large farms are there-fore slightly more vulnerable to climate change than small farms in Latin America. This is likely to be due to large farms specializing in high-value crops and livestock that are both heat-intolerant.

The Latin American analysis also compares the performance of rainfed and irrigated farms. Small and large rainfed farms are expected to lose 4 percent and 30 percent respectively of their farmland value in the PCM scenario. The losses increase to 35 percent and 90 percent with the CCC scenario. In contrast, small and large irrigated farms lose 18 percent and 31 percent of their land value respectively in the PCM scenario, and they lose 31 percent and 48 percent of their land value in the CCC scenario. Irrigated farms are still sensitive to warming but they do not suffer the same losses as the rainfed farms.

A final set of Ricardian impact results concerns China (Wang et al., 2008). A mild 2.5°C warming with an increase of 8 percent precipitation would lead to a small gain of 4 percent of net revenue across all farms. A severe 5°C warming with no increase in precipitation would lead to a 4 percent reduction in net revenue. Examining just rainfed farms, the mild scenario would lead to an 8 percent reduction in net revenue, whereas the severe scenario would lead to a 35 percent reduction in net revenue. With irrigated farms, both the mild and severe scenario would lead to about a 25 percent gain in net revenue. The heavy reliance on irrigation makes China's agriculture less vulnerable to warming. Of course, this robustness is dependent upon there being enough water to continue irrigating farms.

ADAPTATIONS

The cross-sectional analysis of farm choices also provides important insights into adaptation, as shown in Chapter 8. For example, models of crop choice explain how climate affects which crops farmers choose to grow (Kurukulasuriya and Mendelsohn, 2008b; Seo and Mendelsohn, 2008c). Climate clearly affects the selection of crops in both Africa and Latin America. In Africa, farmers in dryer locations choose maize, millet and groundnuts, whereas they grow cowpea and sorghum in wetter loca-tions. In cooler locations, African farmers grow cowpea and beans but in warmer locations they switch to maize, millet and sorghum. In Latin America, farmers chose wheat and potatoes in cooler locations but fruits and vegetables in warmer locations. In dryer locations, farmers chose maize, soybeans and potatoes, and in wetter places they chose rice and fruits and vegetables.

Another key set of decisions that farmers make concerns livestock. They must decide whether or not to own livestock, which species to own, and the desirable size of the herd of that species. In Africa, farmers tend to raise livestock in warmer and dryer locations (Seo and Mendelsohn, 2006). With warmer temperatures, farmers are less likely to choose beef cattle or chickens, but more likely to choose goats and sheep. With less precipitation, they are more likely to choose beef, chickens and goats but less likely to choose dairy cattle and sheep. With warmer temperatures or less precipitation, African farmers choose larger herds of sheep, goats and dairy cattle but smaller herds of beef cattle and chickens. Beef respond poorly to warmer temperatures because at least the commercial beef tends to be heat intolerant. Cattle and sheep may be responding to climate largely through the effect of climate on ecosystems. Climates that lead to good forage such as savannahs will tend to be more attractive to cattle and sheep, whereas goats and chickens can do quite well in forests. Latin America livestock choice is also sensitive to climate (Seo and Mendelsohn, 2007). Global warming will be likely to cause a reduction in beef cattle and pigs in Latin America but an increase in sheep.

Farmers must also make irrigation decisions on their farms. The choice to irrigate is clearly sensitive to the availability of water. But the choice is also a function of climate. In Africa, farmers are more likely to irrigate in dryer and warmer locations (Kurukulasuriya and Mendelsohn, 2008c).

IMPACTS AND ADAPTATIONS

A final set of results mixes the findings of the adaptation research with the measurements of impacts. These structural Ricardian papers model how farmers choose farm types, irrigation, crops or livestock, and then model the conditional incomes earned from each choice. The result is an expected income or land value. By comparing the expected outcome of current versus future climates, one can measure the impact of climate change. Unlike the traditional Ricardian approach, however, this approach also indicates how the farmer has adapted.

The results for irrigation in Africa indicate that with wet and mild future climate scenarios such as PCM, farmers will be able to increase irrigation substantially across Africa and reap large benefits (Kurukulasuriya and

Mendelsohn, 2008c). In contrast, with a hot and dry scenario such as CCC, irrigation will contract slightly and there will be damages.

With African crop farmers, PCM will lead to an increase in millet, groundnut, cowpea, and especially fruits and vegetables but a decrease in maize and sorghum (Kurukulasuriya and Mendelsohn, 2008b). The change in expected crop incomes leads to a 49 percent gain. With the CCC scenario, farmers will switch away from maize, groundnut and sorghum and towards cowpea, fruits and vegetables, and millet. The change in expected income is a loss of 5 percent of income. Interestingly, if the farmers were not allowed to switch crops, there would be a 68 percent predicted loss in income in the CCC scenario.

The results for livestock in Africa indicate that the PCM scenario will reduce beef cattle, dairy cattle, sheep and chickens but increase goats (Seo and Mendelsohn, 2008d). The PCM climate scenario will lead to an expected gain of 6 percent of livestock net revenue. The CCC scenario will lead farmers away from choosing dairy, goats and chickens and toward selecting sheep. The CCC climate scenario will lead to an expected 168 percent increase in livestock net revenue.

In Latin America farmers will switch farm types with global warming (Mendelsohn and Seo, 2007). In the PCM scenario, they will move to crop-only rainfed farming and mixed farming and away from irrigated and livestock farming. This will lead to a 50 percent gain in expected farmland value. But in the CCC scenario, farmers will switch from mixed and especially crop-only farming into livestock-only farming. This will lead to a 17 percent loss in expected farmland value. Note that these predictions are much less severe than the traditional Ricardian predictions of losses of 11 percent to 20 percent in the PCM scenario and losses of 49 percent to 69 percent in the CCC scenario for small and large farms respectively (Seo and Mendelsohn, 2008b).

CONCLUSIONS

The results indicate that agriculture is sensitive to climate. Impacts vary depending on the original climate of a farm and the change in climate. Farms in cooler locations are expected to benefit from warming, whereas farms in hotter locations will clearly be harmed. Changes in precipitation are also important. Farms in very dry locations must either have irrigation or resort to low-value crops that are drought-tolerant. Farms in wetter locations have the luxury of relying on rainfed farming methods.

Although it has been hypothesized that small farms are more sensitive to climate change than large farms, the evidence is mixed. Large farms can be more climate-sensitive than small farms because they specialize in crops and animals that may be heat-intolerant, varieties and breeds designed for temperate climates. Large farms may actually have fewer substitutes as climate warms that would earn a similar level of income. Household farms, in contrast, rely more readily on local crops and species that are better adapted to local conditions. These small farmers have more choices that would earn similar levels of income across different climate scenarios.

The sensitivity of farms to climate appears to vary from one place to another. The empirical results from each region are not the same. It is important to conduct empirical studies of climate impacts in each region of the world. This range of outcomes is partly due to the range of climates in every location. But it may also be due to other factors such as the availability of irrigation, soils and other factors. The results strongly suggest that irrigated farms are less vulnerable to climate change than rainfed farms. Provided there is enough water for irrigation, irrigated farms are much less sensitive to warmer temperatures. Countries with more irrigation appear to be less sensitive to climate than countries with less. Soils can also affect climate sensitivity; for example, some soils are better able to hold soil moisture than others and thus reduce the risks from drought.

Farms may also have different climate sensitivities because of their capital endowments and levels of technology. Farms in more developed countries earn higher net revenues per hectare. However, the evidence is still weak regarding whether capital or technical change alters the climate sensitivity of farms. Surprisingly, farm incomes and farmland values appear to have nothing to do with the objective characteristics of the farmers themselves. Gender, age, experience and education all play an insignificant role in farm net revenues.

Another insight of the climate agriculture literature is that it is important to examine more than just staples. Although staples might represent individual crops of great importance, they reflect only a small fraction of the entire agriculture sector. They are also not representative. Staples (with the exception of rice) tend to be grown in temperate climates. There are many additional crops such as fruits and vegetables that prefer warmer locations. Examining just maize, soybeans, wheat and rice can give a misleading impression of the sensitivity of the entire crop sector. Further, livestock is an important component of the agricultural sector. In warmer and dryer locations, farmers will substitute livestock for crops. By including the full suite of agricultural activities, one can get a more representative impression of how the sector will fare if climate changes.

REFERENCES

IPCC (Intergovernmental Panel on Climate Change) (2007a), *Climate Change 2007: The Physical Science Basis*, Fourth Assessment Report, Cambridge, UK: Cambridge University Press.

IPCC (Intergovernmental Panel on Climate Change) (2007b), *Climate Change 2007: Impacts, Adaptation and Vulnerability*, Fourth Assessment Report, Cambridge, UK: Cambridge University Press.

Kurukulasuriya, P. and R. Mendelsohn (2008a), 'A Ricardian analysis of the impact of climate change on African cropland', *African Journal of Agriculture and Resource Economics*, **2**, 1–23.

Kurukulasuriya, P. and R. Mendelsohn (2008b), 'Crop switching as an adaptation strategy to climate change', *African Journal of Agriculture and Resource Economics*, **2**, 105–26.

Kurukulasuriya, P. and R. Mendelsohn (2008c), 'Modeling endogenous irrigation: the impact of climate change on farmers in Africa', World Bank Policy Research Working Paper 4278, Washington, DC, USA: World Bank.

Mendelsohn, R. and A. Dinar (1999), 'Climate change, agriculture, and developing countries: does adaptation matter?', *The World Bank Research Observer*, **14**, 277–93.

Mendelsohn, R. and A. Dinar (2003), 'Climate, water, and agriculture', *Land Economics*, **79**, 328–41.

Mendelsohn, R. and M. Reinsborough (2007), 'A Ricardian analysis of US and Canadian farmland', *Climatic Change*, **81**(1), 9–17.

Mendelsohn, R. and S.N. Seo (2007), 'An integrated farm model of crops and livestock: modeling Latin American agricultural impacts and adaptation to climate change', World Bank Policy Research Working Paper 4161, Washington, DC, USA: World Bank.

Mendelsohn, R., A. Dinar and A. Sanghi (2001), 'The effect of development on the climate sensitivity of agriculture', *Environment and Development Economics*, **6**, 85–101.

Mendelsohn, R., W. Nordhaus and D. Shaw (1994), 'Measuring the impact of global warming on agriculture', *American Economic Review*, **84**, 753–71.

Parry, Martin, Cynthia Rosenzweig and Matthew Livermore (2005), 'Climate change, global food supply and risk of hunger', *Philosophical Transactions of the Royal Society B*, **360**, 2125–38, published online, 24 October.

Reilly, J., W. Baethgen, F. Chege, S. van de Geijn, L. Erda, A. Iglesias, G. Kenny, D. Patterson, J. Rogasik, R. Rotter, C. Rosenzweig, W. Somboek and J. Westbrook (1996), 'Agriculture in a changing climate: impacts and adaptations', in R.T. Watson, M.C. Zinyowera and R.H. Moss (eds), *Climate Change 1995: Intergovernmental Panel on Climate Change Impacts, Adaptations, and Mitigation of Climate Change*, Cambridge, UK: Cambridge University Press.

Rosenzweig, C. and M. Parry (1994), 'Potential impact of climate change on world food supply', *Nature*, **367**, 133–38.

Sanghi, A. and R. Mendelsohn (2008), 'The impacts of global warming on farmers in Brazil and India', *Global Environmental Change*, **18**(4), 655–65.

Schlenker, W., W.M. Hanemann and A.C. Fisher (2005), 'Will US agriculture really benefit from global warming? Accounting for irrigation in the hedonic approach', *American Economic Review*, **95**(1), 395–406.

Seo, S.N. and R. Mendelsohn (2006), 'Climate change adaptation in Africa: a microeconomic analysis of livestock choice', World Bank Policy Research Working Paper 4277, Washington, DC, USA: World Bank.

Seo, S.N. and R. Mendelsohn (2007), 'An analysis of livestock choice: adapting to climate change in Latin American farms', World Bank Policy Research Working Paper 4164, Washington, DC, USA: World Bank.

Seo, S.N. and R. Mendelsohn (2008a), 'Climate change impacts and adaptations on animal husbandry in Africa', *African Journal of Agriculture and Resource Economics*, **2**, 65–82.

Seo, S.N. and R. Mendelsohn (2008b), 'A Ricardian analysis of the impact of climate change on South American farms', *Chilean Journal of Agricultural Research*, **68**(1), 69–79.

Seo, S.N. and R. Mendelsohn (2008c), 'An analysis of crop choice: adapting to climate change in Latin American farms', *Ecological Economics*, **67**(1), 109–16.

Seo, S.N. and R. Mendelsohn (2008d), 'Measuring impacts and adaptation to climate change: a Structural Ricardian model of African livestock management', *Agricultural Economics*, **38**, 150–65.

Wang, J., R. Mendelsohn, A. Dinar, J. Huang, S. Rozelle and L. Zhang (2008), 'Can China continue feeding itself?: The impact of climate change on agriculture', World Bank Policy Research Working Paper 4470, Washington, DC, USA: World Bank.

12. Policy implications and future research needs

It is now well understood that climate change will not have an identical impact on everyone. Some nations and individuals may be affected more than others, depending on geographical location, level of wealth, infrastructure development and institutional capacity. In the case of agriculture, adaptation plays a critical role as a key proactive measure for coping with likely impacts. In turn, adequate policy is a prerequisite for successful preparedness. We start with the scientific background of climate change, and ask what policy makers need to know and take into consideration in order to make more resilient decisions.

GENERAL NATURAL SCIENCE AND ECONOMIC LINKS TO POLICY

Climate change affects the performance of both plants and livestock. The change in crop and animal performance in turn affects economic outcomes on the farm. Scientific and economic evidence provides critical background information for policy makers to incorporate in their design of intervention policies.

Crops

Greenhouse gases are expected to have many effects on plants, both directly and indirectly. Plants directly respond to the levels of CO_2 concentrations in the atmosphere. Plants also directly respond to changes in climate caused by greenhouse gases. But crops may also be affected by indirect changes such as changes in water availability for irrigation, changes in pests, and changes in weeds. Scientific inquiries that quantify the links between CO_2 concentrations and crop performance, yields and climate, changes in water availability, new levels of pests and weeds are all quite important to policy makers.

Elevated CO_2 concentrations allow plants to reduce the openness of their stomata, leading to a reduction in the water requirements of plants. This tends to increase the water efficiency of crops by reducing the amount of water plants need for the same level of production. Irrigation systems would consequently not need as much water per hectare. However, higher CO_2 levels could also lead to greater leaf area, which would increase plant water consumption. In addition, the stomata closure leads to an increase in leaf temperature since less water is evaporated at the leaf surface. As a result, water vapor pressure increases, leading to an increase in transpiration. The combined effect of all of these factors may limit the overall water saving effects from higher CO_2 concentrations.

As discussed in Chapter 2, elevated CO_2 levels are expected to increase crop yields. However, the magnitude of the effect is greater in C4 than in C3 crops (USDA and USCCSP, 2008). C4 species (for example, maize, sorghum, sugar cane, millet) respond better than C3 plants (for example, cotton, rice, wheat, barley, soybeans, sunflower, potatoes, most leguminous and woody plants) to higher CO_2 concentrations. Because of these different effects in C3 and C4 crops, the fertilization effect of CO_2 will vary with the cropping patterns (mixture of C3 and C4 crops) in each region. The magnitude of the carbon fertilization may vary depending on other limiting factors including nutrients, soil quality and diseases. There is consequently still some uncertainty about what level of carbon fertilization will occur in each region.

The combination of increased CO_2, temperature and humidity can increase both the level of activity and the range of pests. Climate change may increase the ratio of carbohydrate to nitrates in leaves, leading to interactions between plants and herbivores and predators that are the consumers of the crops' organs. The same effects may be observed with weeds (most of which are C4 plants), which can be more competitive than crop plants in certain circumstances. Therefore, policies aimed at growing vulnerable crops would need to incorporate such indirect consideration of yield effects.

The changes in yields and water efficiency will have repercussions on farm economic outcomes. If farmers make no adjustments, reductions (or increases) in yields will lead to proportional reductions (increases) in gross revenues. However, as discussed in Chapters 9 and 10, economic studies reveal that farmers will make adjustments that lead to higher net revenues. They will move away from crops that are no longer productive towards crops suited for the new climate in their local area. These adjustments, or adaptations, must be taken into account, or policy makers will overestimate the damages from climate change. Policies need to encourage profitable adjustments to be taken.

Livestock

Science and economics have made major contributions to the management of livestock under different climate conditions. As discussed in Chapter 2, livestock are affected by the thermal environment – effective ambient temperature and radiation humidity, air movement and precipitation. The performance of many animals is quite sensitive to the climate in which they live, especially if they live outdoors. Climate also has indirect effects on livestock through impacts on the local ecosystem. Greater precipitation moves land from deserts to grassland (savannah) to forests. These indirect effects are very important to livestock, which depend on the local ecosystem for food. Finally, farmers can help livestock adapt by adjusting food intake, metabolism and heat dissipation. For example, dairy cattle raised in barns can be sprayed with water to help them stay cool. However, these latter adaptations are only really possible when animals are raised in shelters and food is bought on the market. As discussed in Chapter 3, it is therefore no surprise that climate has a greater impact on livestock in developing countries (where animals are raised outdoors) compared to livestock in developed countries (where livestock are often raised in sheltered feedlots).

As with crops, farmers can make many adjustments in response to climate. Even if farmers are committed to raising livestock outdoors off the land, there are still many adaptation strategies available to farmers. For example, farmers can switch between raising crops and livestock. Livestock are more suited to dryer climates. Farmers can switch species; they can move from species that are suited to temperate climates to species that can prosper in warmer climates. Further, farmers can switch breeds within a species. They can also adjust the size of their herds as climate conditions change and the productivity of local lands are affected. These adjustments affect the final economic outcomes on each farm. Policies that facilitate these profitable adjustments can thus make a significant difference.

DEVELOPING VS. DEVELOPED COUNTRIES AND REGIONAL VS. NATIONAL POLICIES

This book promotes several methods of measuring climate impacts and adaptations. As discussed in Chapters 7, 8, 9 and 10, these methods have been applied to many countries in various continents over a period of several years. Each of these countries face a wide range of climatic conditions and different possible future climates. The work was also conducted across countries at various levels of development. The policy implications we take away from these works are very comprehensive and detailed.

Developing vs. Developed Country Policies

One lesson for policy makers is that adaptation policies must be tailored to local conditions because the local impacts of climate change vary a great deal across space. Policy makers must be careful transferring interventions from one country to another to make sure that they are appropriate in each place. Technologies, management practices, and crop or livestock varieties that were proven successful in one country need to be carefully evaluated in order to be introduced in another country and still remain successful as climate changes. Institutions, level of infrastructure development, and human capacity, are still factors that may support or prevent good practices from being extrapolated from one country to another.

It is expected that developed (northern hemisphere) countries may actually benefit from climate change, and if negatively impacted, the magnitude would be mild, at most. The cool initial temperatures, existing cropping pattern mix, the level of technology, and the market and processing infrastructure, make many developed country agricultural sectors almost climate-neutral. Therefore, policies to support adaptation to climate change in developed countries such as the USA may not be appropriate for developing countries. Agriculture in developed (mid and high latitude) countries is much less affected by climate change than agriculture in developing (low latitude) countries. A small increase in temperature will cause a lot more damage to developing country farms compared to developed country farms. The specific changes that developed countries make may not be the changes that are most appropriate for developing countries. For example, as temperatures warm, some cooler developed countries may want to move towards a crop that is ideal for moderately warm temperatures, whereas a moderately warm developing country may want to move away from that same crop.

In some circumstances, developing countries may benefit from policies designed and implemented in developed countries. Best management practices and new technologies may be transferable from developed to developing countries. Irrigation technologies, crop varieties and fertilizer applications created in developed countries could be modified by the Consultative Group on International Agricultural Research (CGIAR) system to conditions in developing countries. They then could be carefully introduced by having full support systems (such as agricultural extension, finance, services) in place.

Although each country may need to determine the specific adjustments to climate change that are most appropriate for local conditions, there is one lesson from developed countries that should be transferred to developing countries. Developed countries have prospered because of their reliance on strong property rights for market activities. Strong private property rights are also likely to be important to climate adaptation. In order for farmers to pursue adaptation on their own, they need to have secure ownership of their land and water. Improving private property rights increases the incentive for local farmers to make long-term adjustments (investments) to climate change. It makes no difference where a farm is located for this to be effective. Secure private property rights are especially important in developing countries where vast natural resources are still held in public trust or as common property. To the extent that the best use of these resources is in market activities such as raising crops and livestock, it is important that these lands be privatized. Farmers with private rights will make the investments in adaptation that farmers with insecure rights will not (Deininger and Jin, 2006).

Local vs. National Policies

In large countries such as India and Brazil, the climate, landscape, institutions and local capacities may vary a great deal across the country. Such countries need to be careful to design different policies for different regions within the country. Even smaller countries must be careful not to rely on uniform policies unless every region is similar. Our studies of India and Brazil, as well as the United States and China (Chapter 8), suggest large variations in climate, economic conditions, and other factors across the country. Similar results were found in our continental studies of Africa and Latin America. This range of circumstances calls for a range of responses tied to each local condition.

One possible way to address physical and structural differences in a country is to develop different policies depending on the set of agro-climatic conditions and the existing infrastructure and institutions in the various regions in the country. The reported studies in this book suggest that one would need to have regional policies that fit each region. For example, a recent set of studies reported in Chapter 9 (Kurukulasuriya and Mendelsohn, 2008a, 2008b; Seo and Mendelsohn, 2008b, 2008c, 2008d, 2008e; Seo et al., 2008a, 2008b, 2008c, 2008d) reveal that farmers in each location should have a different response to climate change. Further, the economic and institutional ability to implement adaptation measures may also vary. It is possible that farmers facing similar climate situations may be affected differently, depending on other physical and economic/institutional conditions they face. Both physical and economic/institutional conditions may affect the type of adaptation relevant for each location and the ability of the farmers residing in each location to adapt. Therefore, policy makers should consider tools that tailor assistance as needed. Policy makers may do a better job if they look carefully at impact assessments to identify the most attractive adaptation options. They should apply policies across the landscape using a 'quilt' rather than a 'blanket' approach. The proposed quilt approach will allow the required spatial flexibility and will be likely to lead to much more effective and locally beneficial outcomes.

POLICY REGARDING LARGE VS. SMALL FARMS

The various analyses reported in Chapter 8 provide mixed evidence regarding whether small and large farms differ in their vulnerability to climate change. Large farms have more capital and more advanced technology. Having access to knowledge and alternative adaptation options also makes commercial farms less vulnerable to climate change. However, large farms are also more specialized in high-value temperate farming methods. This makes them more vulnerable to warming. The empirical evidence comparing the performance of small and large farms across climates was therefore mixed. Large African livestock farmers were much more vulnerable than small farmers to warming. Large farms in Latin America, however, were slightly more vulnerable to warming than small farmers. One cannot conclude that small farms are always more vulnerable compared to large farms.

Policies designed for small versus large-scale farms may need to address different aspects of farm operation that are critical for each type of farm. In the following sections we address the support policies that may be critical for small and large farms, such as agricultural extension, microfinance, research and development, insurance, infrastructure, and research and development.

POLICY REGARDING CROPS, LIVESTOCK, AND INTEGRATION OF CROPS AND LIVESTOCK

The analyses reported in Chapter 9 and Chapter 10 suggest that farmers can respond to climate change by introducing many adaptations. Farmers change their decisions over whether to grow crops or livestock, crop mix, livestock mix, and irrigation or rainfed agriculture, depending on climate. These adaptation measures are quite important as they affect the economic outcomes of climate change. Policy makers need to examine whether they can effectively support farmers by promoting changes that help farmers adapt.

As discussed in Chapter 9 and Chapter 10, there is strong evidence that farmers will adapt to climate change by altering farm type. This adaptation may be quite complex, depending on the climate scenario and the location of each farm. Farmers can also adapt to climate change by adding or removing livestock from their portfolio. They can also switch species. These adaptations vary across regions. The distribution of crops across regions in the future will be different depending upon which climate scenario occurs. Some crops will change a great deal across the landscape, such as wheat, fruits and vegetables, whereas others may be more robust. For example, wheat and maize may decline in developing countries but cotton and fruits and vegetables may increase. Other crops, such as rice and sugar, may remain stable. Policy makers should take note of the spatial variation of desired adaptations across the regions.

First, adaptation policies must fit local conditions. Policy makers must be careful not to apply changes broadly across disparate local areas. In general, climate adaptation policies need to take a quilt design to the landscape rather than a blanket design, making sure each patch fits local conditions. Second, adaptations must be designed to match local climate changes. If rain increases in a local area, local farming must adjust to more precipitation, even if rain declines elsewhere in the region. Global predictions of climate change are not particularly relevant. What are needed are accurate local measures of current climate. Third, only broad institutional changes such as improving the efficiency of public water management or increasing private property rights should be applied universally. Such institutional changes create the incentive for all farmers to make long-term adaptation investments. Creating a setting in which farmers are inclined to make the right choices will be beneficial across every country.

IRRIGATION DEVELOPMENT POLICIES

Irrigation is likely to be an important adaptation option, where water is available. Our analyses that compare irrigated versus rainfed agriculture find that the value (or net revenue) per ha of irrigated land surpasses that of rainfed land. In most cases, the differences are large enough to justify investment in irrigation. For example, in the US, the availability of surface water increases the value of farms significantly (Mendelsohn and Dinar, 2003). Dropping irrigated farms from the analysis increases dramatically the temperature sensitivity of US farms (Schlenker et al., 2005). These results suggest that rainfed farms are far more sensitive to warming than irrigated farms. If irrigated farms could replace rainfed farms, the system would be less climate-sensitive (subject to the availability of water). The analysis also examined the effectiveness of various irrigation technologies Gravity irrigation takes advantage of natural potential energy for convey-ance and so is much less expensive, but it also requires more water per ha. Drip irrigation is more expensive but it conserves water. As water becomes more scarce, increased investments into drip irrigation can allow more land to be switched from rainfed to irrigated crops. The process essentially substitutes capital for climate.

The quantitative results for Africa (Kurukulasuriya et al., 2006; Kurukulasuriya and Mendelsohn, 2008a, 2008b) provide direct meas-ures of the relative benefit of rainfed versus irrigated cropland. The study finds sizable differences in the value of irrigated and rainfed agriculture; it further finds that irrigated cropland is much less sensitive to climate than rainfed cropland. These results again suggest that irrigation could help alle-viate the likely effects of climate change. Where water is available, moving from rainfed to irrigated agriculture would increase not only average net revenue per hectare but also the resilience of agriculture to climate change. Governments could make public investments in infrastructure and canals for rainwater harvesting, water storage and conveyance where appropriate and where the public good nature of these investments prevents adequate private sector investment. Policy makers may want to consider support-ing such coping interventions for climate change, where appropriate. The importance of water also suggests that governments need to pay more attention to how water is actually used. As scarce water becomes the con-straint to adaptation, governments need to allocate water more efficiently.

The study of agriculture in China (Wang et al., 2008) further supports the results in Africa and the US. Irrigated farms are more valuable per ha and they are more climate-resilient. In fact, the study suggested that the marginal impact of temperature on irrigated lands was actually positive, whereas warming was predicted to be harmful to rainfed lands. These results stress the importance of irrigation to China's agriculture system. With nearly 60 percent of its cultivated land currently irrigated, China is already very dependent on irrigation. Further, this strong dependence on irrigation suggests China will be able to cope with near-term climate change. How well China will be able to cope with future warming sce-narios depends, in large part, on irrigation. Future water availability will be critical for China's agriculture if climate warming makes water increas-ingly scarce. China may have to look at large investments to transfer water from water abundant to water scarce regions such as the Yellow-to-Hai inter-basin transfer that is now under implementation. However, such engineering solutions may not be sufficient or cost effective in the long run if they are not accompanied by effective water management policies. China will also need to improve its conjunctive use of groundwater and surface water and its allocation of water across users if it wishes to have sufficient water in the future to support agriculture.

SUPPORT POLICIES: RESEARCH, EXTENSION, FINANCE, INSURANCE, INSTITUTIONAL STRENGTHENING AND INFRASTRUCTURE DEVELOPMENT

In addition to the many changes that farmers can make for themselves, governments can also make significant contributions to adaptation. First, they can conduct research and development that leads to new crops and animals that are more suited for hotter and possibly dryer conditions. Although the private sector is effectively developing new high-value plants and animals for farms in temperate climates, they are not investing in new choices for farms that are located in more marginal growing conditions. Governments could encourage the development of varieties with more tolerance for the hot and dry conditions that climate change may bring. For example, Brazil developed a new soybean variety that was well suited for the hotter and dryer climate of the Mato Grasso. Although the purpose of this new variety was to extend farming into a new area, it proves that it is possible to make crops suitable for a warmer climate. Even climate-neutral technical advances will help farmers increase productivity and counterbalance losses from climate change.

Governments can also help strengthen outreach and dissemination programs that provide farmers with advice about alternatives more suitable for changing climates. By carefully training extension officers about new profitable alternatives, they can help farmers adjust quickly to changing climates as they unfold. The agents can also share weather information to make sure that farmers are aware of changes in weather in their region. This will help farmers to be aware of how climate is actually changing in their local area.

Governments can provide credit to help farmers invest in their land and farming operations. Governments can create and protect private property rights so that farmers have the incentive to adapt autonomously. One problem that is often been cited for small farmers is that they do not have access to formal sources of credit. Creating micro lending facilities can address this problem on a local level.

Governments can provide better access to reduce the cost of farmers getting their product to market. They can invest in roads and railroads that reduce the cost of transport from remote farms to cities and ports. Governments can invest in better port facilities so they can both import and export crops at low cost. Governments can encourage international trade in crops to help reduce local dependence on crop productivity. This is especially important given the likely changes in crop and livestock mixes at the local level.

Governments can encourage economies to develop and diversify away from agriculture so that only a small fraction of developing country economies are at risk from climate change. Although there is no question that agriculture is very sensitive to climate change, most economic sectors are not sensitive. To the extent that developing countries can reduce the share of their economy in agriculture, they can substantially reduce their overall vulnerability to climate.

Finally, falling local productivity may make some areas unable to support current population levels. Governments should help people migrate internally within their countries to regions with more promising economic opportunities. For example, governments can help growing metropolitan regions plan for more growth. Governments must also resist the temptation to subsidize people staying in places that are no longer viable.

In addition to encouraging direct adaptations, local and national governments and international organizations could invest in infrastructure and institutions to ensure a stable environment to enable agriculture to prosper. Such policy interventions may not only achieve the long-term goal of helping vulnerable populations adapt to climate change, but they may also increase the likelihood of achieving the more immediate Millennium Development Goals, such as halving hunger, reducing poverty and improving health.

FUTURE RESEARCH

The analyses in this book have provided many insights into global warming impacts on agriculture, but many questions still remain unanswered. While there has been a great deal of progress in understanding the economics of climate change, there is still more to be explored. What is still unresolved? We provide a list of both technical issues and policy issues that still need to be resolved.

First, the Ricardian technique needs further research. What is the appropriate functional form of the Ricardian function? Should the dependent variable be net revenue or farmland value? The choice in past studies was largely determined by what was available. What should the measure be in the future? Should the dependent variable be net revenue per ha or the log of net revenue per ha? Many past studies have used net revenue, but the log of net revenue may be a more appropriate choice. How should climate be characterized? The initial characterization of climate used four separate months to capture seasons. Subsequent studies used the average of three months to capture each of four seasons. Yet other studies used a single index of heating degree days in the growing season. What is the best description of climate in every country? What other independent variables should be included in future regressions as control variables? What has been omitted that should be included? Finally, how can future technologies be accounted for in the Ricardian model? The current models capture how farmers use current technology. Is there any way to modify these models to capture future technology as well?

One interesting result across the different studies is that every continent has different coefficients. Partly this may be due to the availability of data, as no two studies have exactly the same data. However, the African and South American studies have very similar data and still they lead to different results. Should the Ricardian method develop a single model that can be estimated everywhere or should different models be estimated in different continents or countries? The Ricardian climate coefficients also change over time. Does this reflect a weakness in the technique or are there missing variables that would explain this shift if included? What variables are changing over time that would explain the observed shifts? All these questions suggest that future additional work is important and needed.

Another area that calls for more research is the interface between agriculture and water. If one of the possible responses to global warming is a greater reliance on irrigation, and if climate change continues to reduce available water supplies, water is going to be an increasingly limiting factor. In order to address this issue, it is critical that a set of new studies be developed that model both water and agriculture together. Although there are some examples of such studies in developed countries, they are very rare in developing countries.

This book discusses the direct effect of climate on agriculture. However, the book assumes that prices or goods and inputs remain the same. It is likely, however, that at least in some local areas, the prices of some services will be affected by climate change. For example, if certain rural areas become unproductive, the demand for laborers will decline and local wages will fall. The book does not discuss the likely outmigration that will result, nor the changes in the local economy that will follow. The book does not discuss how changes in local production will eventually alter the flows of international trade. However, the analyses in the book did recognize that regional trade agreements between countries could be important.

Another potentially important interaction that is not examined is between health and agriculture. Health can affect agriculture if it reduces local labor supply. Without adequate manpower, some areas will simply not be farmed. Agriculture, in turn, affects health. Without sufficient food or income to buy food, people cannot stay healthy. Climate can affect this interaction through its effect on farm productivity. But it is also true that climate can affect this interaction through its affect on health. If climate change increases the likelihood that people fall ill in an area, it may affect the viability of farms there. The Ricardian method captures this interaction given current government policies. Farms that are in warm climates are being compared to farms that are in cooler places given all the diseases and handicaps of each place. However, if governments addressed a health threat by, for example, promoting public health measures to control a disease, that could change the relationship between climate and disease. It is therefore important that this link between agriculture and health be studied more closely.

Another interesting research topic concerns the use of agro-ecological zones to forecast climate impacts (Kurukulasuriya and Mendelsohn, 2008a). Agro-ecological zones may help researchers determine how changing climates will affect different local areas. It may be especially useful for extrapolating from a limited sample to a diverse geography (Seo et al., 2008a, 2008b, 2008c, 2008d; Seo and Mendelsohn, 2008e). Further, it may be possible to forecast how these zones will shift over time and to use this information to predict how cropping patterns and climate impacts will change (Kurukulasuriya and Mendelsohn, 2008). This agro-ecological zone approach sheds a different light on how farmers adapt.

REFERENCES

Deininger, K. and S. Jin (2006), 'Tenure security and land-related investment: evidence from Ethiopia', *European Economic Review*, **50**(5), 1245–77.

Kurukulasuriya, P., R. Mendelsohn, R. Hassan, J. Benhin, T. Deressa, M. Diop, H.M. Eid, K.Y. Fosu, G. Gbetibouo, S. Jain, A. Mahamadou, R. Mano, J. Kabubo-Mariara, S. El-Marsafawy, E. Molua, S. Ouda, M. Ouedraogo, I. Sène, D. Maddison, S.N. Seo and A. Dinar (2006), 'Will African agriculture survive climate change?', *The World Bank Economic Review*, **20**(3), 367–88.

Kurukulasuriya, P. and R. Mendelsohn (2008a), 'How will climate change shift agro-ecological zones and impact African agriculture?', World Bank Policy Research Working Paper 4717, Washington, DC, USA: World Bank.

Kurukulasuriya, P. and R. Mendelsohn (2008b), 'Crop switching as an adaptation strategy to climate change', *African Journal of Agriculture and Resource Economics*, **2**, 105–26.

Kurukulasuriya, P. and R. Mendelsohn (2008c), 'A Ricardian analysis of the impact of climate change on African cropland', *African Journal of Agriculture and Resource Economics*, **2**, 1–23.

Mendelsohn, R. and A. Dinar (2003), 'Climate, water and agriculture', *Land Economics*, **79**(3), 328–41.

Schlenker, W., W.M. Hanemann and A.C. Fisher (2005), 'Will US agriculture really benefit from global warming? Accounting for irrigation in the hedonic approach', *American Economic Review*, **95**(1), 395–406.

Seo, S.N. and R. Mendelsohn (2008a), 'A Ricardian analysis of the impact of climate change on South American farms', *Chilean Journal of Agricultural Research*, **68**(1), 69–79.

Seo, S.N. and R. Mendelsohn (2008b), 'Measuring impacts and adaptation to climate change: a structural Ricardian model of African livestock management', *Agricultural Economics*, **38**, 150–65.

Seo, S.N. and R. Mendelsohn (2008c), 'An analysis of crop choice: adapting to climate change in Latin American farms', *Ecological Economics*, **67**, 109–16.

Seo, S.N. and R. Mendelsohn (2008d), 'Climate change impacts and adaptations on animal husbandry in Africa', *African Journal of Agriculture and Resource Economics*, **2**, 65–82.

Seo, S.N. and R. Mendelsohn (2008e), 'A structural Ricardian analysis of climate change impacts and adaptations in African agriculture', World Bank Working Paper 4603, Washington, DC, USA: World Bank.

Seo, S.N., R. Mendelsohn, P. Kurukulasuriya and A. Dinar (2008a), 'An analysis of adaptation to climate change in African livestock management by agro-ecological zones', *B.E. Journal of Economic Analysis & Policy*, (forthcoming).

Seo, S.N., R. Mendelsohn, A. Dinar, R. Hassan and P. Kurukulasuriya (2008b), 'A Ricardian analysis of the distribution of climate change impacts on agriculture across agro-ecological zones in Africa', *Environmental and Resource Economics*, (forthcoming).

Seo, S.N., R. Mendelsohn, A. Dinar, R. Hassan and P. Kurukulasuriya (2008c), 'Differential adaptation strategies to climate change in African cropland by agro-ecological zones', World Bank Policy Research Working Paper 4600, Washington, DC, USA: World Bank.

Seo, S.N., R. Mendelsohn, A. Dinar, R. Hassan and P. Kurukulasuriya (2008d), 'Long-term adaptation: selecting farm types across agro-ecological zones in Africa', World Bank Policy Research Working Paper 4602, Washington, DC, USA: World Bank.

United States Department of Agriculture (USDA) and US Climate Change Science Program (USCCSP) (2008), *The Effects of Climate Change on Agriculture, Land Resources, Water Resources, and Biodiversity*, Washington, DC: USDA–ARS.

Wang, J., R. Mendelsohn, A. Dinar, J. Huang, S. Rozzelle and L. Zhang (2008), 'Can China continue feeding itself? The impact of climate change on agriculture', World Bank Policy Research Working Paper 4470, Washington, DC, USA: World Bank.

Index